PSYCHOLOGY LED ASTRAY

PSYCHOLOGY LED ASTRAY
CARGO CULT IN SCIENCE AND THERAPY

TOMASZ WITKOWSKI

BrownWalker Press
Boca Raton

Psychology Led Astray: Cargo Cult in Science and Therapy

Copyright © 2016 Tomasz Witkowski
All rights reserved.

All rights reserved. No part of this book may be reproduced or transmitted in any form or by any means, electronic or mechanical, including photocopying, recording, or by any information storage and retrieval system, without written permission from the publisher.

BrownWalker Press
Boca Raton, Florida • USA
2016

ISBN-10: 1-62734-609-0
ISBN-13: 978-1-62734-609-2

www.brownwalker.com

Cover art by Pieter Bruegel the Elder [Public domain],
via Wikimedia Commons

Publisher's Cataloging-in-Publication Data

Names: Witkowski, Tomasz.
Title: Psychology led astray : cargo cult in science and therapy / Tomasz Witkowski.
Description: Boca Raton, FL : BrownWalker Press, 2016. | Includes bibliographical references.
Identifiers: LCCN 2016937894 | ISBN 978-1-62734-609-2 (pbk.) | ISBN 978-1-62734-610-8 (PDF ebook)
Subjects: LCSH: Child psychotherapy. | Psychotherapy. | Psychology. | Mental health. | Pseudoscience. | MESH: Psychotherapy. | Psychology, Child. | BISAC: PSYCHOLOGY / Psychotherapy / Counseling. | PSYCHOLOGY / Applied Psychology. | PSYCHOLOGY / Psychotherapy / Child & Adolescent.
Classification: LCC RC480 .W556 2016 (print) | LCC RC480 (ebook) | DDC 616.89/14--dc23.

Let them alone; they are blind guides.
And if the blind lead the blind, both will fall into a pit.
 – Matthew 15:14

Table of Contents

Foreword IX
Introduction XI

Part I: Is Psychology a Cargo Cult Science? 15
Chapter 1: From the Pacific Ocean to Social Sciences:
How the Cargo Cult has Reached Science 17
Chapter 2: From Cargo to Voodoo: How Stress Kills 33
Chapter 3: Self Reports and Finger Movements of the Weirdest
People in the World 49
Chapter 4: Ignorance, Nonchalance, or Conspiracy?
Actual Statistical Significance of Research Results 61
Chapter 5: Neuroscience: From Phrenology to Brain Porn 77
Chapter 6: Why Has Science Transformed into a Cult? 95

Part II: Uncontrolled Experiments on Humans:
Cargo Cult in Psychotherapy of Adults 107
Chapter 7: Adult Children of Alcoholics or Victims of the
Barnum Effect? 113
Chapter 8: Between Life and Death: The Simonton Method 133
Chapter 9: The Disaster Industry and Trauma Tourism:
The Harmful Effects of the Psychological Debriefing 149
Chapter 10: Experimental Therapy Patient's Handbook 165

Part III: Is There Anything You Could Not Do for Your Child? 173
Chapter 11: A Saint or a Charlatan?
The Doman-Delacato Method 175
Chapter 12: Go to School, Get a Taste of Pseudoscience:
Educational Kinesiology 191
Chapter 13: Warning! Hugging can be Dangerous to Life
and Health! Attachment Therapy 201
Chapter 14: Dolphins: Wonder Therapists, Intelligent Pets,
or Aggressive Predators? 217
Chapter 15: Ad Infinitum 229
Chapter 16: Why is This All so Common? 241
Chapter 17: Protecting Yourself From Charlatans 261

Dear Richard 267

Foreword

This is a very important and valuable book, both for the social scientist who values rigourous scientific inquiry and for the layperson interested in the state of modern psychology. Its value lies in part because Tomasz Witkowski has summarized so succinctly and so powerfully the many problems that plague modern psychological research and its application, and in part because he has done so in such an engaging manner. Although dealing with material that could by its nature constitute a dry and boring read, he has enlivened his critical commentary by weaving through it entertaining and informative strands of storytelling, both literary and historical.

But this is also likely to be a disturbing book for that same social scientist who values rigourous scientific inquiry, as well as for the layperson who may at some point seek psychological assistance. It is disturbing because it holds a mirror up to psychological research and practice, and much of what is reflected is not very pretty. Witkowski begins with reference to physicist Richard Feynman's characterization of social science as a cargo cult, and he makes a compelling case that is there is much about modern psychology that fits with that pejorative analogy: He points to the persistence of inappropriate and misleading statistical practices that underlie much of psychological research, practices that have endured despite decades of criticism and condemnation. He addresses the essentially parochial nature of much of psychological research that, for the most part, has been conducted with participants who are representative of only a small segment of the populations of a few nations that in turn represent only a small segment of the world at large. He skewers the many fads and crazes that seem to sweep regularly through applied psychology (and in some cases, that stop sweeping and become firmly entrenched). He targets the proliferation of untested varieties of psychotherapy; the overgeneralization and exaggeration of findings from neuroscience; the creation from whole cloth of "disorders" such as the Adult Children of Alcoholics syndrome in the absence of any appropriate evidence; the continuing use of discredited treatment approaches such as Trauma Debriefing, Attachment/holding therapy, Facilitated Communication, and Dolphin Therapy. And for the layperson, he augments his critical examination of psychological therapies with guidance for anyone seeking psychotherapy.

However, Witkowski does not throw out the baby with the bathwater; he concludes his penetrating criticism by challenging Feynman's "cargo cult" appraisal and pointing out that evidence-based psychology has made very significant contributions to both understanding and improving the human condition.

This is a well-written book, and the extensive documentation that is provided makes it an excellent source of reference material as well. Every psychologist, every psychology student, and every layperson with an interest in psychology or with need for its services will benefit immensely from reading it. And although it is largely devoted to decrying the egregious departures from scientific rigour that afflict much of our discipline and to skewering the many false and exploitative therapies and techniques that are huckstered to a hungry public, it is ultimately a book that honours and defends science-based psychology. I highly recommend it.

James E. Alcock

Introduction

Our predecessors, in fighting to survive in an environment populated by threats to their safety, were unceasingly engaged in assessing the level of danger to their lives. In making these calculations they were at risk of fulfilling two fundamental errors – either underestimating a given threat, or overestimating it. Most likely, those who too frequently committed the error of disregard simply died out, together with the gene pool partially responsible for such mistakes. So, were the survivors those who rationally judged the level of danger? Unfortunately, in making precise estimates one runs the risk of error, and sometimes even a small mistake is enough to exclude one from the contest where the victor's spoils include passing genes on to the next generation. It is therefore most likely that our forbearers were not necessarily rational individuals, but rather those who overestimated threats, as this was not associated with any particularly serious costs. Going out of the way to walk around thickets of brush where a suspicious rustling could be heard did not demand a lot of energy, whereas skepticism of such suspicions could cost one's life. Avoiding unfamiliar animals and fear of them were more adaptive than an unhealthy curiosity, which could have wound up leading to a fatal bite. Not by accident did evolution leave nearly all of us with a deep-rooted fear of all snakes, even though there are very few poisonous breeds. Keeping our distance from snakes comes at no great cost, and protects us from the few of them whose venom would deny us the chance to remain among the living.

In this same manner, we inherited a fear of unknown secretions and scents, as well as of individuals whose skin is distinctly different from ours, whether as a result of sores, markings, injuries, or even color. In the course of evolution we also cemented our capacity to perceive cause-and-effect relations wherever possible, and to prevent the potential effects of these perceived rules – it does not cost us much to refrain from walking around a cemetery at night, to stay away from deceptive flickers of light over bogs, to spit over our left shoulder three times when a black cat crosses our path, to leave a small offering for elves and spirits both good and bad, to keep our fingers crossed, to keep our guard up on Friday the 13th, and to engage in other such activities. On the other hand, insufficient caution can lead to the most terrible of consequences. Is it, therefore, worth risking? A skeptical attitude could turn out to be a costly one.

Errors of underestimation are of particular significance in the case of identifying intentional beings. The predator which has taken us for its dinner, the enemy of the same species that is lurking and hunting for our things, our lives; this is one of the most serious threats which we have faced in the course of our species' evolution. This explains why so many irrational beliefs are of a personal nature, or assume the existence of intentional beings. This is why we have filled the forests with gnomes and pixies, witches, werewolves, and warlocks; cemeteries, ruins, old homesteads, and castles are filled with cursed souls, ghosts, nightmares, and vampires; rivers, lakes, and swamps have been left to nymphs, ogres, and nixes; earthquakes, storms, floods, and droughts are ascribed to the gods.

The mind of the *Homo sapiens* species, which has spent at least 200,000 years exercising its ability to identify associations and dependencies, as well as intentional beings where they have never existed, is fertile soil for irrationalism and can sustain all of its forms. Critical thought is an unnatural act for such a mind, and skepticism is a dangerous attitude, yet the circumstances in which it evolved lost their significance long ago. This also concerns the minds of scholars, who, aware of their limitations, have developed a social system of control in the form of science. Unfortunately, this system is full of holes like Swiss cheese, as we have demonstrated in *Psychology Gone Wrong: The Dark Sides of Science and Therapy*.[1]

The thicket of ideas existing today, whose authors battle for the largest possible number of pliable minds, is in no way reminiscent of the Pleistocene savannah where the human mind was formed. While failure to spot an intentional being stalking us from the bushes might once have finished tragically, today we may meet an equally tragic end through faith in imagined dependencies, superficial associations, and nonexistent influences. Some people pay for these convictions with their life, while others rush off into the fog of promises only to lose their time and money; there are also those who succumb to contemporary illusions and are deprived of their freedom, their children, and their fortune. These barbarian practices take place all around us, and are frequently shielded only by a thin, opaque layer of tolerance, elevated to the status of the highest value. We imitate refinement through acceptance, and afford political correctness to indifference for the activities of idiots and fraudsters. We have adopted laws which help us send people to prison for offending our religious sensibilities, regardless of how absurd these feelings are and of the suffering their implementation brings to others. At the same time, we permit the widespread insulting of rational understanding.

Psychology, a very young and immature science, engaged in one of the most complex pieces of subject matter that can be encountered in the universe, is particularly exposed to all of these problems I have written about above. Its representatives are descendants of Pleistocenes, which is why their minds are also prone to the mistakes common of *Homo*. For these reasons, this book has been written and dedicated to demasking the statements which, in the field of psychology, under the cover of promises and good intentions, hide mistaken beliefs, lies, the desire for profit, and sometimes even blind cruelty.

In the first part I examine the well-known metaphor which was applied in 1974 at a commencement speech for students of the California Institute of Technology given by Richard Feynman. This genius physicist compared the contemporary social sciences to a cargo cult. Was he justified in doing so?

Are the social sciences, including psychology, really one massive illusion similar to the ones constructed by natives living on the islands of the Pacific Ocean, waiting forlornly for planes that will bring them goods packed in boxes marked cargo? And if not, what differentiates the practices of cargo from that which we can call science? The book's title already gives a hint of the answer, but I am all too aware that it is not entirely obvious.

In the second part I achieved two goals. On the one hand, I wanted to use some trends in therapy to demonstrate how psychology has also become home to the cargo cult. The selection of examples does not, however, serve only to fit into Feynman's metaphor. The task, no less important, which I set for myself in that section was to demask successive therapies devoid of scientific bases; therapies with dangerous, or even cruel consequences. And while I succeeded in describing merely a few of the hundreds of pseudotherapies on the market, I hope that the knowledge presented in those chapters will save some readers from fruitless searches. The last chapter was titled "How to Protect Yourself? The Patient's Guide to Experimental Therapies," and it contains tips helpful in making a good decision when selecting a therapy, as well as information useful in difficult situations that arise during the course of therapy.

The third section concerns child therapy, an exceptional field. While mistakes in choosing a therapy for adults or engagement in pseudoscientific experiments with oneself playing the lead can be dismissed by declaring "adults can decide what they want to do and they have the right to do it, because they pay and they suffer the consequences of their own decisions", the selection of influences on a child and the costs that child will incur should not be at the parents' sole discretion; this is particularly true when the child's health or life is at risk. As in the preceding part, this section concludes with a guide, this time for parents, titled "Protecting Yourself From Charlatans." In it, I present helpful tips to be applied during the difficult decision of selecting a therapy for your own child.

I know that my book will not be an easy read, neither for scientists nor for countless therapists, or for those who have wasted time and resources on ineffective therapies. Many of them, I am sure, will remain unconvinced. But since prevention is the best cure, please treat my book as rather a vaccine than a treatment – may it be a chance for all those who have not yet been sucked in by pseudoscience.

As Richard Feynman said, "So we really ought to look into theories that don't work, and science that isn't science."[2] This appeal has long been forgotten, and what is being done in the name of psychology, the field I practice every day, goes to show that it was really never heard. Let us, then, take a serious look into science that isn't science.

---

[1] T. Witkowski and M. Zatonski, *Psychology Gone Wrong: The Dark Sides of Science and Therapy* (Boca Raton: Brow Walker Press, 2015).

[2] R. P. Feynman, "Cargo Cult Science," *Engineering and Science 37*, (1974): 10-13.

PART I:
IS PSYCHOLOGY A CARGO CULT SCIENCE?

Chapter 1: From the Pacific Ocean to Social Sciences: How the Cargo Cult has Reached Science

The beginning of the 1930s brought fear and horror to the native inhabitants of the New Guinea Mountains. There, lo and behold, giant birds appeared in the sky and made a terrifying and deafening roar. At the very sight of those monsters flying above them, the natives fell down to the ground in terror, shielded their faces with their hands, and prayed for mercy. That phenomenon occurred so unexpectedly and was so utterly different from anything they had ever known, not only from their own personal experience, but also from their legends and stories, that they quickly got the idea that the birds must have been the enraged spirits of their ancestors. The growls of those beasts were more chilling than the rumble of an earthquake, the thud of a waterfall, or any sound made by any animal known to them. They were in no doubt that the monstrous clamor could not have been made by any terrestrial being. With the intention of alleviating such ancestral wrath, the natives decided to sacrifice their most valuable possessions – they slaughtered their pigs, roasted them, and scattered pieces of meat all over the fields as propitiatory food for their forefathers. They then picked up croton leaves and rubbed them all over their hair and backs in order to protect themselves from sinister forces.

With time, the "great birds from heaven," as the locals called them, grew in number and the people could see the bizarre creatures landing on their ground more frequently. The islanders observed that the birds' giant "bellies" opened up and released groups of white visitors who carried such things as tools, weapons, food supplies, and other goods from their depths. The locals watched these scenes unfolding before their eyes in deep silence and filled with anxiety. Hundreds of questions were racing around in their heads. Where do these people come from? Why are only their faces and hands uncovered? Are they white because they have descended from heaven where it is closer to the Sun? Can they conjure supernatural forces? Are they our ancestral spirits?

Naturally, the void conceived by such questions could not remain unaddressed. The initial fear gradually yielded to certainty; certainty that the "great birds from heaven" sent by their forefathers would also soon visit them, the islanders, and bring them similar goods, all packed in similar boxes labelled cargo. The people just needed to be worthy and deserving of such gifts, which they would be if they began to venerate the gods by performing rituals and making appropriate sacrifices. Soon the islanders became convinced that the rituals alone would not be a sufficient means to coax their deities, and thus they extended their activities beyond their traditional ceremonies. Observation of the white men bustling around in a manner that bore no resemblance to any activity they had known led the locals to believe that what they were seeing must be some unusual religious rites aimed at enticing "the great birds from heaven" into landing. Filled with pious zeal, they started to emulate certain actions taken by the white people. The locals built more and more huts specifically designed to accommodate the cargo that was eventually going to be delivered to them. They constructed full-sized fake airplanes made of wood and cane – because, of course, these "great birds" were in fact airplanes – and offered sacrifices to them. At night they built bonfires to serve as landing beacons for the "birds" to see where to land, and if some aircraft unexpectedly came into their view, they fell to the ground in devotion, waiting anxiously for cargo to come...

This is just one of the abounding examples of how the cargo cult was formed on the Pacific Ocean islands. The referenced cult of the great bird from heaven developed in the early 1930s, mainly among native tribes inhabiting the mountain ranges of New Guinea. At that time, the civil administration commenced a regular, aerial reconnaissance of the island's interior.[3] The reported activities of the local tribes were not a unique occurrence. Similar cults flourished on several dozen other Pacific islands. However, they were not always centered upon the airplane motive; sometimes the natives awaited trucks or ships or, in other cases, the multiplication of money was supposed to take place. Depending on the local social context and conditions, the cults took diverse forms. The most captivating were – and still are, since the cargo cults have remained alive even until today – far and away the ones related to the faith that the airplanes were birds sent from heaven. Let us take this opportunity to take a closer look at some of them.

In 1943, the Morobe province on New Guinea in the Markham River valley saw its inhabitants build huge "radio houses" and install "telephone lines." It all started with and resulted directly from their observation of the behavior of Japanese soldiers who had occupied the Huon peninsula since early 1942. The troops stationed in that area used telephone communications in their operations, for which temporary telephone lines were needed. Let us quote an eyewitness account of the effect this had on locals:

> The ANGAU officers who were exploring the area noticed that in the villages of Arau and Wompur, bamboo poles had been placed in "radio shacks," with "wires" (ropes) conducted first onto the roof to bamboo "insulators" and then – as if a type of an overhead system – into the adjoining "radio shack." Each morning military drills were held with bamboo canes used as rifles. When the islanders were later asked why they were doing that, they answered that they were getting ready for the coming of Christ and the phone installation would allow them to receive the message about such Coming. ...
> When there appeared fighting Allied and Japanese planes in the sky, the expectant villagers assumed it was their forefathers struggling with the Whites who had yearned to prevent the ancestors from coming and bringing cargo whatever the cost. As a consequence of that judgment, prayers became more intense and more sacrifices were offered to gods. Bonfires were built around the "observation towers" to illuminate the landing site to the ancestors. ... When the waiting

time prolonged and neither the forefathers nor cargo arrived, the islanders decided to improve their "telephone system" – they got down to replacing "insulators" and "cables" (ropes) and fitting them in different locations.[4]

As part of the *skin guria* cult prospering on New Guinea in the area of Pindiu, mock airfields were also set up near cemeteries for cargo planes that they assumed would reach those places. Moreover, the islanders expected trucks and ships loaded with goods. They were believed to arrive from America through a hole dug deep in the ground. Houses were built in the bush to be used as storage facilities for the much awaited cargo.

Followers of the Kaum cult, originating largely from the Garia tribe, likewise constructed runways in the belief that their forefathers would send cargo by air so that it would not be stolen by European marauders. The locals also made bonfires at nights and sent out light signals with lamps so the spirits could find their way in the dark and land safely on the well-prepared landing sites.

Spiritual leaders of the Peli Association, which in fact represents an extremely modern cult since it emerged as recently as the 1960s, foretold their followers that 300 American 707 Boeings would land on the very top of the Turu mountain on July 7th, 1971, all with cargo and large amounts of money onboard. Two days before the revealed date, several thousand people set off for the mountain in order to participate in the delivery of this wealth. Sadly, the prophecy was not fulfilled and the religious cult members returned to their homes in silence.[5]

Cargo cults were not without influence on the islanders' lives. Events that took place in the 1930s on New Guinea within the Marafi cult community illustrate well the dramatic impact they made on people's daily practices: "In anticipation of the arrival of their ancestors who had been believed to bring rice, meat, food and material goods, the people ceased their work and abandoned their fields and gardens. They were blindly obedient to Marafi's orders, even when he demanded that girls and women be given to him as his wives."[6]

One of the prophecies formulated by John Frum – a legendary religious leader from the island of Tanna in the New Hebrides Archipelago – put the locals on the verge of economic collapse: "John Frum also prophesied that, on his second coming, he would bring a new coinage, stamped with the image of a coconut. The people must therefore get rid of all their money of the white man's currency. In 1941 this led to a wild spending spree; the people stopped working and the island's economy was seriously damaged."[7]

Yet another example is that of members of the Kaum cult on New Guinea, who concluded that they would only be deserving of a reward from the gods if they first destroyed all their belongings: "Gardens were being destroyed, pigs slaughtered and family possessions burnt just to show loyalty to the Kaum leader's orders and to appear poor and humble in the gods' eyes so as to make them send cargo."[8]

In their religious zeal, the cargo cult followers burnt entire villages to the ground and even committed murder if it was decided that a human sacrifice was necessary to coax the gods. Despite such large-scale offerings, the cargo never came. At that point, the followers would sometimes double their efforts; they would pray yet more fervently and make even greater sacrifices. Occasionally, doubts were raised as to the credibility of the prophecies, and then the cult naturally ceased to exist. Many a time, people turned against the religious leaders themselves and, accusing them of fraud and abuse, imprisoned them. Some were eventually expelled by the very same people who had hitherto been their most passionate followers. Nevertheless, there were also cases where, after the local administration forbade some practices, the indigenous population would come to the conclusion that they were close to reaching their objective and would practice their rituals with increased determination and devotion. Cults that were dying out were replaced by new ones that revived unfulfilled hopes for material abundance. Some cargo cults have managed to linger on until this day. Islanders unfailingly build their runways and bonfires. People gather on their "airfields," around their meticulously built "radio stations," and on "control towers," continuing to look hopefully at the sky, everyone silent and focused.

The issue at hand for us is thus: how did it come to pass that an extraordinary phenomenon akin to the cargo cult could have taken root in the social sciences?

In 1974, during a graduation ceremony (the Caltech 1974 commencement address) held at the California Institute of Technology, Richard Feynman, a physics genius and Nobel prize winner, gave an exquisite speech entitled "Cargo Cult Science. Some Remarks on Science, Pseudoscience and Learning how to not Fool Yourself." Its key element was a highly symbolic comparison that left a lasting effect on the minds of people of science:

> I think the educational and psychological studies I mentioned are examples of what I would like to call cargo cult science. In the South Seas there is a cargo cult of people. During the war they saw airplanes with lots of good materials, and they want the same thing to happen now. So they've arranged to make things like runways, to put fires along the sides of the runways, to make a wooden hut for a man to sit in, with two wooden pieces on his head to headphones and bars of bamboo sticking out like antennas – he's the controller – and they wait for the airplanes to land. They're doing everything right. The form is perfect. It looks exactly the way it looked before. But it doesn't work. No airplanes land. So I call these things cargo cult science, because they follow all the apparent precepts and forms of scientific investigation, but they're missing something essential, because the planes don't land.[9]

The great physicist did not restrict himself to this mere comparison. To support his metaphorical claim, Feynman provided examples in the areas of healing, psychotherapy, and parapsychology. He also related his experiences with isolation tanks (sensory deprivation chambers) and made observations on education and crime:

> So I found things that even more people believe, such as that we have some knowledge of how to educate. There are big schools of reading methods and mathematics methods, and so forth, but if you notice, you'll see the reading scores keep going down – or hardly going up – in spite of the fact that we continually use these same people to improve the methods. There's a witch doctor remedy that doesn't work. It ought to be looked into; how do they know that their method should work? Another example is how to treat criminals. We obviously have made no progress – lots of theory, but no progress – in decreasing the amount of crime by the method that we use to handle criminals.
> Yet these things are said to be scientific. We study them.[10]

Other critics of the social sciences share Feynman's viewpoint. In point of fact, in his book *Social Sciences as Sorcery* Stanislav Andreski does not refer to practices in the social sciences as a cargo cult, but he explores the very same phenomenon:

> When a profession supplies services based on well-founded knowledge we should find a perceptible positive connection between the number of practitioners in relation to the population and the results achieved. Thus, in country which has an abundance of telecommunication engineers, the provision of telephonic facilities will normally be better than in a country which has only a few specialists of this kind. The levels of morality will be lower in countries or regions where there are many doctors and nurses than in places where they are few and far between. Accounts will be more generally and efficiently kept in countries with many trained accountants than where they are scarce. We could go on multiplying examples, but the foregoing suffice to establish the point.
> And now, what are the benefits produced by sociology and psychology? ... So to examine the validity of the claim that these are highly useful branches of knowledge, let us ask what their contribution to mankind's welfare is supposed to be. To judge by the cues from training courses and textbooks, the practical usefulness of psychology consists of helping people to find their niche in society, to adapt themselves to it painlessly, and to dwell therein contentedly and in harmony with their companions. So, we should find that in countries, regions, institutions or sectors where the services of psychologists are widely used, families are more enduring, bonds between the spouses, siblings, parents and children stronger and warmer, relations between colleagues more harmonious, the treatment of recipients of aid better, vandals, criminals and drug addicts fewer, than in places or groups which do not avail themselves of the psychologists' skills. On this basis we could infer that the blessed country of harmony and peace is of course the United States; and that ought to have been becoming more and more

so during the last quarter of the century in step with the growth in numbers of sociologists, psychologists and political scientists.[11]

Where Feynman concludes that there is no connection between efforts taken by representatives of the social sciences and the condition of the field within their main interest, Andreski takes the point further. He puts the blame on the scientists themselves for the fact that some problems continuously grow: "It may be objected that this is no argument, that the causation went the other way round, with the increase in drug addiction, crime, divorce, race riots and other social ills creating the demand for more healers. Maybe; but even accepting this view, it would still appear that the flood of therapists has produced no improvement. What, however, suggests that they may be stimulating rather than curing the sickness is that the acceleration in the growth of their numbers began before the upturn in the curves of crime and drug addiction."[12]

Both Feynman and Andreski look into reasons why the discipline is in such an alarming condition. According to the former, the transformation of science into a cargo cult has taken place mostly due to the lack of honesty on the part of investigators themselves. "Honesty" here should be understood in a particular manner:

> It's a kind of scientific integrity, a principle of scientific thought that corresponds to a kind of utter honesty – a kind of leaning over backwards. For example, if you're doing an experiment, you should report everything that you think might make it invalid – not only what you think is right about it: other causes that could possibly explain your results; and things you thought of that you've eliminated by some other experiment, and how they worked – to make sure the other fellow can tell they have been eliminated.
> Details that could throw doubt on your interpretation must be given, if you know them. You must do the best you can – if you know anything at all wrong, or possibly wrong – to explain it.[13]

At the same time, Feynman cautions scientists against yielding to the temptation of self-deception: "The first principle is that you must not fool yourself – and you are the easiest person to fool. So you have to be very careful about that. After you've not fooled yourself, it's easy not to fool other scientists. You just have to be honest in a conventional way after that."[14]

The failure of scientists to maintain integrity towards themselves, their peers, and those outside science represents, in Feynman's view, the underlying reason for science turning into a cargo cult. In his study, Andreski makes a more radical point. As a sociologist by education, he argues that distortions of reality stem from social factors.

> Though formidable enough, the methodological difficulties appear trivial in comparison with the fundamental obstacles to the development of an exact sci-

ence of society which puts it on an entirely different plane from the natural sciences: namely the fact that human beings react to what is said about them. More than that of his colleagues in the natural sciences, the position of an "expert" in the study of human behaviour resembles that of a sorcerer who can make the crops come up or the rain fall by uttering an incantation. And because the facts with which he deals are seldom verifiable, his customers are able to demand to be told what they like to hear, and will punish the uncooperative soothsayer who insists on saying what they would rather not know – as the princes used to punish the court physicians for failing to cure them. Moreover, as people want to achieve their ends by influencing others, they will always try to cajole, bully or bribe the witch-doctor into using his powers for their benefit and uttering the needed incantation … or at least telling them something pleasing. And why should he resist threats or temptations when in his specialty it is difficult to prove or disapprove anything, that he can with impunity indulge his fancy, pander to his listeners' loves and hates or even peddle conscious lies. His dilemma, however, stems from the difficulty of retracing his steps; because very soon he passes the point of no return after which it becomes too painful to admit that he has wasted years pursuing chimeras, let alone to confess that he has been talking advantage of the public's gullibility. So, to allay his gnawing doubts, anxieties and guilt, he is compelled to take the line of least resistance by spinning more and more intricate webs of fiction and falsehood, while paying ever more ardent lip-service to the ideals of objectivity and the pursuit of truth.[15]

Contrary to Feynman, who sees practitioners of the social sciences as victims of self-deception, Andreski builds up an image of a cynical, obsequious scientist who has a personal interest in deforming reality.

The easiest way out is always not to worry unduly about the truth, and to tell people what they want to hear, while the secret of success is to be able to guess what it is that they want to hear at the given time and place. Possessing only a very approximate and tentative knowledge, mostly of the rule-of-thumb kind, and yet able to exert much influence through his utterances, a practitioner of the social sciences often resembles a witch-doctor who speaks with a view to the effects his words may have rather than to their factual correctness; and then invents fables to support what he said, and to justify his position in the society.[16]

Other critics of the social sciences are of the same mind as Feynman and Andreski. Marvin Minsky states unequivocally that to this day psychology has been unable to develop tools that would allow understanding of the nature of thinking processes or consciousness.[17] Many authors disapprove of the jargon of social sciences, which they claim is frequently employed to make terms sound more complicated than is necessary. By using such overblown verbosity some simply try to give the impression that they have something wise to say.[18] Karl Popper ruthlessly reproaches those scientists for such practices: "Every intellectual has a very special responsibility. He has the privilege and opportunity of studying. In return, he owes it to his fellow men (or 'to society') to represent the results of his study as simply, clearly and modestly as he can. The worst thing that intellectuals can do – the cardinal sin – is to try to set themselves up as great prophets vis-à-vis their fellow men and to impress them with puzzling philosophies. Anyone who cannot speak simply and clearly should say nothing and continue to work until he can do so."[19]

Later in this part of the book I will make an effort to determine which factors responsible for the transformation of social sciences into a cargo cult have played the most significant roles in the process. However, before I investigate the matter, let me first shed some light on how Feynman's metaphor, though based on observations rather than on research, has begun to live a life of its own. A Google search for the expression "cargo cult science" brings nearly 30,000 hits. In the majority of cases, online texts accessed through the Internet use this term for some kind of pseudoscience or activities that are doomed to fail. Scientific databases, such as EBSCO, also offer a high number of papers that include the "cargo cult" phrase. Just a handful of titles selected at random provide clues as to how the metaphor has been contextualized:

- "Neuro-linguistic programming: Cargo cult psychology?"[20]
- "Cargo cult science, armchair empiricism and the idea of violent conflict."[21]
- "The urban question as cargo cult: Opportunities for a new urban pedagogy."[22]
- "Classroom research and cargo cults."[23]
- "Environmental optimism: Cargo cults in modern society."[24]
- "Dominance theater, slam-a-thon, and cargo cults: Three illustrations of how using conceptual metaphors in qualitative research works."[25]
- "On cargo cults and educational innovation."[26]
- "Cargo-cult city planning."[27]
- "Psychology – 'a cargo cult science'? In search of developmental psychology paradigm."[28]

It seems that Feynman's metaphor has already established itself in the public domain. And though I value Feynman and his perspective on science, I decided to test to what extent his famous comparison was justified by the reality of the times it comes from, and also whether his observations would be confirmed today. Here is the outcome of my investigation.

To start with, in Feynman's times the United States saw a 300% rise in the number of serious crimes, such as murders, rapes, or armed robberies, committed per capita (comparing data from 1960 and 1986).[29] One can imagine how much research was conducted on crime, rehabilitation, and other related issues at that time, how many articles were published, and how many people pursued an academic career investigating these aspects of social life. So Feynman gets a point for that.

Statistics on teenage suicides in the United States were similarly discouraging. Between 1950 and 1980, the number of suicides in this age bracket also increased by 300%. The highest rise was noted among white teenagers from better-off social classes. For instance, 1985 alone saw a total of 29,253 suicides, of which 1,339 were committed by whites aged 15–19.[30] According to the PsycLIT base, at that time 1,642 reviewed articles on suicide were published. This massive research energy combined with tragic statistics seem to support Feynman's view on the social sciences.

Figures showing mental health conditions appear equally gloomy. In 1955, 1.7 million patients were admitted to psychiatric hospitals in the United States, with the number reaching 6.4 million in 1975.[31] This amounts to a nearly four-fold increase. During the same period, thousands of clinical psychologists were engaged in intense efforts to improve methods of therapy, prophylaxis, and other aspects. Ultimately, if their labor had borne fruit, this should have been reflected in the statistics.

As Feynman had predicted, the quality of education deteriorated as well. The 1967 average mathematics SAT test score in high school graduates was 466, and in 1984 it went down to 426. A corresponding decline was reported in verbal skills test scores.[32]

Similar statistics relating to ever-worsening serious problems and pathologies are available in abundance. It is sad to say, but such recitals are true not only for Feynman's times. They also apply to the present day and seem even more shocking today. Many authors, such as Marcia Angel, claim that:

> It seems that Americans are in the midst of a raging epidemic of mental illness, at least as judged by the increase in the numbers treated for it. The tally of those who are so disabled by mental disorders that they qualify for Supplemental Security Income (SSI) or Social Security Disability Insurance (SSDI) increased nearly two and a half times between 1987 and 2007 – from one in 184 Americans to one in seventy-six. For children, the rise is even more startling – a thirty-five-fold increase in the same two decades. Mental illness is now the leading cause of disability in children, well ahead of physical disabilities like cerebral palsy or Down syndrome, for which the federal programs were created.
> A large survey of randomly selected adults, sponsored by the National Institute of Mental Health (NIMH) and conducted between 2001 and 2003, found that an astonishing 46 percent met criteria established by the American Psychiatric Association (APA) for having had at least one mental illness within four broad categories at some time in their lives.[33]

The World Health Organization (WHO) on its website states that in 2012 more than 350 million people suffered from depression,[34] which represented nearly five percent of the entire world population. Of this number, a million committed suicide.[35] These figures have shot up surprisingly fast, as in 2010 this rate was substantially lower and totaled 298 million people, accounting for 4.3% of the global population.[36]

The rise in the number of depression cases is reported simultaneously with an increase in the number of mental illnesses and disorders that affect people today. In 1952, the first edition of the *Diagnostic Statistical Manual of Mental Disorders* (DSM) came out, with descriptions of 106 mental disorders. The second issue, in 1968, already listed 182 types. These first two editions were subjected to fairly harsh criticism from psychiatric diagnosticians as incomplete, inaccurate, and generally disappointing in many other respects. It was only the third edition that earned the title of the "psychiatrists' bible" as it listed 205 disorders, that is, nearly two-and-a-half times more than the first edition. In 1987, a revised version of the third edition extended the list to as many as 292 descriptions. It seemed that the classification process had stabilized, because the 1994 DSM-IV added only a few disorders to the list, thus extending it to up to 297. However, the year 2000 saw the publication of the DSM-IV reviewed edition that offered as many as 365 psychiatric diagnoses! One disorder for each day of the year. This version has remained in force until today because DSM-V, which was published in 2012 and extended the already lengthy list yet again, was condemned by the National Institute of Mental Health (NIMH) for its lack of validity. Consequently, this argumentation led to the withdrawal of financial support of further research conducted by the team of DSM experts.[37] A critical look at the presented figures from outside can give the impression that over the last 60 years mental disorders have gone like a deadly virus through a range of mutations at an alarming rate, and the endeavors of armies of psychologists, psychiatrists, and other mental health care workers could in no way halt the flood of new cases.

A thorough understanding of the situation is possible only if we compare the aforementioned data with statistics that bring us closer to the status of the mental health care profession. "The APA's Center for Workforce Studies estimates that there are 93,000 practicing psychologists in the United States. Licensed psychologists totaled approximately 85,000 in 2004. Graduations average 4,000–5,000 per year and approximately 2,700 of those are in health service provider fields, resulting in an additional 8,100 practicing psychologists."[38]

Interestingly enough, neither the American Psychological Association (APA) nor anyone else in the United States or other countries keeps exact statistics concerning the number of professionally active psychologists in the world, and the data presented are only approximate. One may, however, try to estimate the rate of increase in the number of people who practice this profession. Roy Menninger and John Nemiah, in their book on the post-war history of psychiatry, estimated that between 1974 and 1990 the number of practitioners in the United States grew from 20,000 to 63,000.[39] This constitutes a three-fold growth within barely 16 years.

I have made an attempt at a rough calculation of the growth of the number of psychologists in Wrocław, the city in which I studied. My calculations were based on the number of psychology graduates at the time when I myself graduated from the University of Wrocław and the number of annual graduates that emerge today. If this rate remains stable, there would be more than two million psychologists in Wrocław in just 50 years! This is a number that exceeds the city's population figures by three times. Obviously, such estimates leave out many factors, such as the gradual dying out of the population of psychologists, and therefore they should be treated anecdotally only. More specific information may, however, prove that there is something to it. For instance, the Institute of Psychology at the University of Wrocław nearly doubled its educational capacity with regard to the number of graduates/qualified psychologists, it produces every year. While 119 graduate students received a Master's degree in 2005, in 2009 their number totaled as many as 226!

Similar inferences are drawn from analysis of the growing number of alumni of the social sciences in Poland. According to the *Annual Statistical Yearbook (Mały Rocznik Statystyczny)* in 1998, only 0.04% of Poles graduated from areas of study defined as social sciences. In 2007, this group had increased to 0.16%, which translates into a four-fold growth. The field of psychology made up a large portion of this group. At the same time, the number of patients in mental health clinics doubled.

Based on fairly similar data, it is easy to reach the same conclusions as the one formulated by Andreski years ago: "If the number of psychologists and sociologists continued to grow at the rate it did during the last decade, then it would overtake the total population of the globe within a few hundred years."[40]

An exaggeration? Of course. "A few hundred years" is far too long a period for Andreski's prophecy to come to pass. Anyway, even if the latter statement is slightly far-fetched, such a comparison of the dynamic growth of the number of educated professionals and various figures illustrating a dramatic rise in the number of mental issues and appearance of new mental disorders forces us to give serious consideration to another of Andreski's thoughts: "If we saw whenever a fire brigade comes the flames become even fiercer, we might well begin to wonder what it is that they are squirting, and whether they are not by any chance pouring oil on to the fire."[41]

So are the social sciences, as Feynman argued, cargo cult sciences? Are we involved in an act of profound deception where we bear responsibility for the upkeep of thousands of scientists who engage themselves in utterly idle efforts?

I have frequently participated in discussions on that subject. Voices of concern and outrage raised by psychologists or other representatives of the social sciences by no means surprise me. This perspective calls into question the meaning of their everyday professional activity. I am also very well acquainted with the arguments they present on such occasions. One can hear, for instance: "And what if this is just the increased availability of the mental health care that has exposed actual needs in this field?" Possibly! Let me, however, emphasize that this hypothesis – as it is only a hypothesis – is in principle equal to the assumption that the growth in the number of psychologists and psychotherapists has led to a demand for psychotherapeutic services. Is it not, by any chance, an artificially grown demand? Another line of defense involves the argument that had it not been for psychologists and their endeavors, the number of people suffering from mental issues would have been substantially higher than that shown by statistics today. Indeed, a simple extrapolation of the present trends reveals that within a mere half a century we will all be patients of mental health clinics. Only, who will work there? And perhaps it may already be the case – one might be tempted to ask – that potential patients ignorant of their issues currently provide services to those who have already become aware that they need assistance in this respect.

Other arguments point to the fact that there are too many factors that determine mental health issues for psychologists to be held solely responsible for the present condition. But who can be blamed, if not them? Who has the most extensive knowledge in this regard? Which social group, if not psychologists, should investigate these factors and show how to eliminate their impact? Is this shifting of responsibility onto others not deceptively similar to all the attempts made by the cargo cult priests and followers who frantically seek explanations for the fact that their air fields do not lure any planes to land and unload huge cargo boxes? Let us examine the following account:

> The entire local community lived in the expectation that cargo would come one day and therefore they prayed incessantly, days and nights. Nonetheless, these efforts also proved futile. The failure of cargo to arrive was therefore deemed to have been caused by flaws, be it in the ritual or people. It was consequently agreed that sins must have been committed by the people, and hence absolution was needed. The sins were confessed publicly and then public penance followed, many a time severe and drastic, involving even human sacrifice. Cargo was still absent, however, and so the people tried to employ new means and ritual approaches. In their gardens, they would strip leaves off the trees, root the plants out and eat the fruit, slaughter pigs and hens to make offerings in worship of gods and dead missionaries so as they could contribute to the coming of cargo. A special taboo was established on eating certain foods or having sexual intercourse. And, despite all these efforts, the cargo did not come.[42]

Yet other defenders of today's practices within the social sciences domain may accuse this line of reasoning as being excessively linked with the applied aspect of science. Supporters of this argumentation point to the fact that primary research is not always successful in bringing instantly visible results and sometimes years pass before some discoveries have been put into practice. Furthermore, doubts voiced by Feynman or Andreski are extremely short-sighted as they reduce the work of scientists to merely its practical aspect. Such an exclusively pragmatic perspective on science is alien to me. I am fully aware that many disciplines constitute an arena of intense investigation that will have no or little impact on our reality, or whose impact will be extensively delayed. My expectations of the fields of philosophy, cosmology, or astronomy are not that of groundbreaking discoveries and brisk changes. Instead, I am a fervent advocate of conducting inquiries in these fields. However, individual spheres of science undeniably differ in terms of the degree to which each of them is immersed in the reality it relates to. I find it very difficult to imagine engineers working on the improvement of bridge structures stating that they are in fact doing primary research and it will take some time before results of their work become evident, while subsequent bridges collapse one after another with increasing frequency. The social sciences, including psychology, are very close to reality and even if some of their practitioners undertake primary research, sooner or later their results should manifest themselves.

Undoubtedly, there is a plethora of explanations for the lack of a causal nexus between the number of engaged professionals combined with the amount of effort devoted to problem-solving and the outcomes achieved. It surprisingly rarely – even for a fleeting moment – provides for an interpretation that representatives of the social sciences perhaps err in the process of problem identification. Or maybe they simply ask the wrong questions? Theories that doubt the usefulness of their activities are even rarer. The question arises as to whether this situation, by any chance, suits the representatives of those disciplines?

Refraining from answering these questions for a while, let us undertake a mental experiment; imagine we replace psychotherapists with dental surgeons, and people suffering from mental issues with dental patients accordingly. Imagine that the number of dental professionals grows four-fold in a decade and, at the same time, the number of patients also rises. Initially, we will certainly be persuaded by the fact that an increasing number of patients willingly use professional dental services. However, with time people should visit them only for routine checkups or partly for some cosmetic treatments. Is it possible that with the number of dental surgery professionals increasing so rapidly there will be more and more patients with poor dental health? After all, the high availability of dentists ensures that every tooth issue is handled immediately. It is likely that many out of the myriad of dental clinics would not survive in the market! Nevertheless, in our mental experiment more and more people succumb to tooth decay, periodontitis, or other problems. The dentists explain to us that they do try their best and, at the same time, go on enumerating other factors, such as poor quality food, water, air, and the like, that allegedly cause damage to our teeth. Our collective dental health would have been in much worse condition had it not been for their efforts. Would we give credence to their stories? Would it be unnatural if we expected them to develop more effective methods of tooth protection, prophylaxis, and treatment?

Leaving the social sciences aside, is there any other area where professional services are offered and, similarly, the number of experts or service providing entities grow with a concurrent rise in the number of issues they try to handle? Progress in other fields is not so hard to observe. Our cars consume less and less fuel and become increasingly more efficient. Travelling by air constantly becomes cheaper and far safer. Despite a number of disreputable exceptions, medicine also witnesses ongoing advancements and agricultural agencies struggle but frequently manage to overcome the problem of food shortages in the face of a rising global population (despite futurologists when I was still at school conjecturing that at the threshold of the third millennium mankind would be in danger of at least starvation and privation, if not mired in total disaster). The social sciences indeed seem to be the only disciplines where problems and questions multiply instead of being solved.

Let us now follow Feynman again and look at our civilization through a stranger's eyes. This was done by Feynman at the Galileo Symposium held in Italy in 1964. He tried at Galileo to show the overall condition of scientific culture in modern society and explain why, despite the intensive development of science, it largely remained so utterly non-scientific. If we put ourselves in the mind of a stranger, with our reflections on modern psychology in mind, we would see societies in which the number of people getting into mental trouble is constantly on the rise. This rapid increase in the number of mental issues goes hand in hand with the growing presence of different types of helpers that are mushrooming at a rate exceeding the growth rate of potential patients. Years pass and, paradoxically, issues not only do not vanish, but their number continues to rise. Does this give us the right to affirm the effectiveness of such specialists, or should we rather perceive their activities in the same way as we look at the formation and advancement of cargo cults? Is Feynman perhaps too moderate in this case? Is it rather Andreski's remark about oil being poured on to the fire that is a more appropriate comment in the face of the current picture? Nearly forty years ago, the American author and educator John Holt did not have much doubt in this regard: "The person whose main lifework is helping others needs and must have others who need his help. The helper feeds and thrives on helplessness, creates the helplessness he needs."[43]

In the next chapter I will focus on one particular case to show you how the cargo cult came to exist and how this helplessness is generated.

---

[3] W. Kowalak, *Kulty Cargo na Nowej Gwinei* (Warsaw: Akademia Teologii Katolickiej, 1982), 152-153.
[4] Ibid., 179-180.
[5] Ibid., 226-231.
[6] Ibid., 157.
[7] R. Dawkins, *The God Delusion* (Boston, MA: Houghton Mifflin Harcourt, 2008), 236.
[8] Kowalak, *Kulty Cargo*, 173.
[9] Feynman, "Cargo Cult Science," 11.
[10] Ibid., 10.

[11] S. Andreski, *Social Sciences as Sorcery* (London: Andre Deutsch, 1972), 25-26.
[12] Ibid., 26.
[13] Feynman, "Cargo Cult Science," 11.
[14] Ibid., 12.
[15] Andreski, *Social Sciences*, 24.
[16] Ibid., 31.
[17] M. Minsky, "Smart Machines," in *Third Culture: Beyond the Scientific Revolution*, ed. J. Brockman (New York: Touchstone, 1996), 152-166.
[18] D. Dennet, "Intuition Pumps," in ibid., 181-197.
[19] K. Popper, *In Search of a Better World: Lectures and Essays from Thirty Years* (London: Routledge, 2012), 83.
[20] G. Rodrique-Davies, "Neuro-Linguistic Programming: Cargo Cult Psychology?" *Journal of Applied Research in Higher Education 1*, (2009): 57-63.
[21] B. Korf, "Cargo Cult Science, Armchair Empiricism and the Idea of Violent Conflict," *Third World Quarterly 27*, (2006): 459-476.
[22] R. Shields, "The Urban Question as Cargo Cult: Opportunities for a New Urban Pedagogy," *International Journal of Urban & Regional Research 32*, (2008): 712-718.
[23] E.D. Hirsch Jr., "Classroom Research and Cargo Cults," *Policy Review 115*, (2002): 51.
[24] W.R. Catton Jr., "Environmental Optimism: Cargo Cults in Modern Society," *Sociological Focus 8*, (1975): 27-35.
[25] S. Dexter, D.R., LaMagdeleine, "Dominance Theater, Slam-a-thon, and Cargo Cults: Three Illustrations of how Using Conceptual Metaphors in Qualitative Research Works," *Qualitative Inquiry 8*, (2002): 362.
[26] B. Starnes, "On Cargo Cults and Educational Innovation," *Education Week 17*, (1998): 38.
[27] P. Buchanan, "Cargo-cult City Planning," *Architectural Review 190*, (1992): 4-8.
[28] B.M. Kaja, "Psychologia – 'Nauka Kultu Cargo?' W Poszukiwaniu Paradygmatu Psychologii Wychowania," ["Psychology – is it a 'Science of Cargo Cult?' Searching for a Paradigm of Educational Psychology,"]. *Forum Psychologiczne 11*, (2006) 42-57.
[29] *U.S. Department of Justice's Uniform Crime Reports*, (July 25, 1987): 41. http://archive.org/stream/uniformcrimerepo1986unit/uniformcrimerepo1986unit_d jvu.txt; *Statistical Abstracts* (1985):. 166; The Commerce Departament's *U.S. Social Indicators*, (1980): 235, 241).
[30] *Vital Statistics of the United States*, (1985) U.S. Department of Health and Human Services, 1988), table 8.5.; *Social Indicators*, (1981).
[31] *U.S. Social Indicators*, (1981): 93.
[32] *U.S. Statistical Abstracts*, (1985): 147.
[33] M. Angell, "The Epidemic of Mental Illness: Why?" *The New York Review*, (June 23, 2011): http://www.nybooks.com/articles/archives/2011/jun/23/epidemic-mental-illness-why/
[34] http://www.who.int/mediacentre/factsheets/fs369/en/
[35] http://www.who.int/mediacentre/factsheets/fs369/en/
[36] T. Vos, A. D. Flaxman, M. Naghavi et al., "Years Lived with Disability (YLDs) for 1160 Sequelae of 289 Diseases and Injuries 1990-2010: A Systematic Analysis for the Global Burden of Disease Study 2010," *Lancet 380*, (December 2012): 2163–96.

[37] C. Lane, "The NIMH Withdraws Support for DSM-5," *Psychology Today*, (May 4, 2013): http://www.psychologytoday.com/blog/side-effects/201305/the-nimh-withdraws-support-dsm-5

[38] American Psychological Association, Support Center, "How many practicing psychologists are there in the United States?" http://www.apa.org/support/about/psych/numbers-us.aspx#answer

[39] R. Menninger, J. Nemiah, *American Psychiatry After World War II: 1944–1994.* (Washington, D.C.: American Psychiatric Press, 2000), 136.

[40] Andreski, *Sorcery in Social*, 222.

[41] Ibid., 28.

[42] Kowalak, *Kulty Cargo*, 152.

[43] J. C. Holt, *Escape from Childhood: The Needs and Rights of Children.* (CreateSpace Independent Publishing Platform; 2 edition, 2013), 45.

Chapter 2: From Cargo to Voodoo: How Stress Kills

On another occasion, while my informant resided on the same spot, passing a tauped place[44] one day he saw some fine peaches and kumaras, which he could not resist the temptation of appropriating. On his return home, a native woman, the wife of a Sawyer, requested some of the fruit, which he gave to her, informing her, after she had eaten it, where he obtained it. Suddenly, the basket which she carried dropped from her hands, and she exclaimed, in agony, that the attua[45] of the chief whose sanctuary had been thus profaned would kill her. This occurred in the afternoon, and next day, by twelve o'clock, she was dead.
Another instance, still more remarkable, came under his observation. The son of a chief had been long unwell, and was a great burden to his parents. His father one day upbraided him with trouble which he had thus become to his friends, and said, that as the attua had determined to keep him unwell he had better die at once, and rejoin his friends; mentioning the name of a particular one to go to in the next world. Without uttering a word, the young chief turned on his side, drew his blanket over his head, and in a few hours he was dead.[46]

This is probably one of the first accounts reported to our civilization about the cases of sudden death which would later be referred to as *voodoo death*. Ascribed to "primitive people," they were so astonishing that one of the best physiologists of the 20th century, Walter Bradford Cannon, when writing about them concluded: "The phenomenon is so extraordinary and so foreign to the experience of civilized people that it seems incredible."[47] We, contemporary members of Western civilization, approach similar ideas with bewilderment and are constantly amazed by their power of suggestion that, in extreme cases, may even lead to death. This disbelief is rooted in our confidence that no such thing can ever happen to us and that it concerns other people only. What lies deep inside us is our hidden feeling of superiority over those simple, primitive minds that are so susceptible to manipulation that the fear they experience leads them to self-destruction. From the distant viewpoint of Western civilization, cases of voodoo death take place only in books authored by anthropologists or in programs broadcast by TV stations focusing on the exotic, such as the Discovery Channel or National Geographic. It does not even occur to us that the voodoo experience takes a heavy toll on us every day, but that nobody makes TV programs about it and victims remain unaware that they have been cursed. In the United States alone, not fewer than 20,000 people die as a result of voodoo practices. Shamans who exercise such tremendous powers are psychologists, the priests of a cargo cult that is imperceptibly taking form of a voodoo rite. We have become aware of this quite recently from the research published in 2012 by Abiola Keller and his collaborators.
The study tracked 28,753 adults in the United States for eight years. To be able to do this, the researchers linked data from the 1998 National Health

Interview Survey to the prospective National Death Index (NDI) mortality data through 2006. The participants of this study were asked such questions as: "During the past 12 months, would you say that you experienced a lot of stress, a moderate amount of stress, relatively little stress, or almost no stress at all?" or "During the past 12 months, how much effect has stress had on your health – a lot, some, hardly any, or none?" Keller and her colleagues examined the factors associated with participants' current health status and psychological distress. Next, they used NDI records to find out who out of the investigated group had died. This procedure enabled them to determine the impact of the belief that stress affects health on all-cause mortality.

The results were astonishing. A total of 33.7% of nearly 186 million U.S. adults perceived stress as a factor affecting their health a lot or to some extent. Both higher levels of reported stress and the perception that stress affects health were independently associated with an increased likelihood of actual worse health and mental health outcomes. The interaction between the amount of stress and the perception that stress affects health proved to be so distinct that those who reported a lot of stress and believed that stress impacted their health a lot had a 43% increased risk of premature death. People who experienced a lot of stress but did not view stress as harmful were no more likely to die. In fact, they had the lowest risk of dying of anyone in the study, including people who had relatively little stress.

The researchers also estimated that over the eight years they were tracking deaths, 182,000 Americans died prematurely: not from stress, but from the belief that stress is bad for their health. This number translates into 20,231 deaths a year. If that estimate is correct, that would make believing stress is bad for your health the 15th largest cause of death in the United States, killing more people than skin cancer, AIDS, and homicide.[48] Welcome to the circle of voodoo ritual...

But this was not the only research study to show the deadly effects of the conviction that stress could be harmful to our health. Herman Nabi and his collaborators decided to examine whether the perceived impact of stress on health is in fact associated with adverse health outcomes. They intended to investigate whether individuals who report that stress adversely affects their health are at increased risk of coronary heart disease as compared with those who report that stress has no adverse health impact. The analyses were based on 7,268 men and women with a mean age of 49.5 years from the British Whitehall II cohort study. Over the 18 years of follow-up, there were 352 cases of coronary deaths or first non-fatal myocardial infarction reported in the study. After adjustment for sociodemographic characteristics, participants who reported at baseline that stress had affected their health "a lot" or "extremely" had a 2.12-times higher risk of coronary death or non-fatal myocardial infarction when compared with those who reported no effect of stress on their health. Although this was not a randomized controlled trial but rather a prospective cohort study, the authors concluded that the perception that stress affects health, regardless of perceived stress levels, was associated with an increased risk of coronary heart disease.[49]

The numbers clearly show that thoughtless repetition of the phrase "stress is harmful to your health" could as a matter of fact be fatally dangerous for people who are prone to believing it. How has it come to this, that so many of us have given credence to this curse and how it actually works? To find answers to these questions, we need to take a closer look at the cargo cult priests who have created an object of worship in the guise of stress which, when you believe in it, kills you.

Stress is probably the most commonly used term in the psychological vocabulary. We discuss the occurrence of stress in the workplace and argue about managerial stress or stress at school. We read about the effect stress has on road safety as well as about its damaging influence on decision-making processes. By the same token, we explore the role of stress in the emergence of conflicts, and most of all the severe impact of stress on our health. Whole journals are devoted to stress (e.g. *Anxiety, Stress and Coping, Journal of Stress Management*, etc.). However, there are also many doubts concerning the phenomenon of stress and its existence is questionable.[50] It is time to look at them with the attention and respect they deserve.

The origins of the concept of stress go back to the turn of the 19th century, when Walter Canon discovered that the blood of animals that had repeatedly been exposed to stressful stimuli during experiments always contained adrenaline. This hormone had already been discovered before his research; however, nobody had suspected that it might have any emotional or mental implications. Soon after having made that discovery, Canon formulated his theory of the *fight or flight* reaction and consequently described the mechanism of *homeostasis*. For the first time in the history of science, Canon's research showed that emotion is a phenomenon that takes place not only in our mind, but also in the whole human body.

The fight or flight reaction, as well as the concept of homeostasis, was soon adopted by physician and endocrinologist Hans Selye, who had ineffectively searched for an unknown female hormone in the body of animals. In autopsies on dead animals used for experiments he identified three common symptoms: stomach ulcers, enlarged adrenal glands, and shrunken tissue of the immune system. Selye came to the conclusion that this could have been an effect of the negative stimulus applied to the animals during experiments. Based on his new assumption, Selye instigated a new series of experiments in which he exposed rats to many different aversive stimuli. Regardless of the type of stimuli used, the same effects were observed and a high level of adrenaline was consistently reported. In order to describe this nonspecific reaction, he borrowed a term from metallurgy – *stress*.[51] Drawing on Canon's discoveries and his own observations, in 1956 he published *The Stress of Life* in which he conceptualized the physiology of stress as having two components:

a set of responses which he termed the general adaptation syndrome, and the development of a pathological state from ongoing, unrelieved stress.[52]

Unfortunately, his theory did not meet with the approval of endocrinologists and physiologists. Incorrect categorization of the reaction and the lack of a precise theoretical definition of stress were the most frequent arguments deployed against his concept. Moreover, Selye claimed that a certain quantity of stress is necessary for life, and hence he introduced two subcategories, i.e. *eustress* and *distress*, thus making the entire notion even more complicated. The new differentiation meant that from now on every emotion fell into the definition of stress and therefore could be referred to as stress. This overgeneralization was, however, the cause of subsequent severe criticism of Selye's theory. At the very birth of the concept of stress, even Walter Canon was quite skeptical as to its scientific worth.

Today things are even more confounding. As the Australian psychologist and critic of the notion of stress Serge Doublet writes:

> The various perspectives from which researchers have approached the problem have added to the confusion. As the concept of "stress" came into vogue, many investigators who had been working with a concept which was felt to be related to stress, substituted the word "stress" and continued with their previous line of investigation. This resulted in concepts such as "anxiety", "conflict", "frustration", "emotional disturbance", "trauma", "alienation" and "anomie" being described as "stress" (Cofer & Appley, 1964). What has also complicated matters somewhat has been that often the same terms have been suggested as being a cause of stress. Frustration, worries, fears, anxiety have also been said to be the result of stress. It seems that all these terms could be "stress", cause it and result from it; which would mean that stress causes stress which in turn causes more stress. A literal cascade of stress.[53]

In the face of such massive criticism as experienced by Selye, scientists generally have only two options: either to redouble their efforts and continue the investigation until a theory gains a descriptive and predictive power, or to abandon their research altogether. But it seems that Selye found his own, third way. Despite the hostile criticism, he started to popularize his theory with increased determination and employed a number of marketing-like techniques for the effective promotion of it. For years he gave lectures on stress and its implications, appeared on radio and TV shows, and also tried to interest physicians, military psychiatrists, and the general public in his concept.

With time, his strategy appeared to be very successful. While it was Cannon who had introduced the notion into scientific use in physiology,[54] Selye managed to make others believe that he was the first one to put it into a biomedical context[55] and today he is commonly recognized as its author. Selye even became known as Dr. Stress, and was nominated for the Nobel Prize ten times. Nowadays, even though the term "stress" is commonly and widely used not only by the public but also by scientists, almost every emotional reaction can be alarmingly easily understood and described as a symptom of stress. Limitations and flaws in his theory that had been repeatedly pointed out by his adversaries have never been addressed. What is even worse, the claim that stress has a negative influence on our health in spite of many controversies in this field has become a common dogma.

How was this possible? Well, as Serge Doublet wrote: "Stress is a wonderful blanket that can be thrown over any problem."[56] While few, if any, know the whole truth about stress, we are prone to blame it for many of our misfortunes. If you have forgotten to take your car papers with you, stress is available to blame. A poor examination score may also be easily ascribed to stress, in the same way as a car collision that you have caused. Apart from excusing one-off events, we are also inclined to hold stress responsible when it comes to our habitual behavior. And so stress may drive us to compulsive overeating and becoming overweight, as well as addictions. Stress is readily raised as the underlying cause of things we do not fully understand, in particular for unexpected health problems that emerge for no apparent reason, from hair loss to cancer. A dentist advised me once that the tooth issues I was struggling with at that time had been most likely caused by stress. The press regularly publishes articles about the economic costs of the stress we are exposed to. If one common cause is used to explain so many different circumstances, it should give rise to our concern as to the accuracy of such explanations. However, since in no way can the accuracy of all such statements be analyzed here, I shall focus my attention on those that are perhaps the most important, that is the ones related to our health.[57]

The ability to properly identify the causes of our injuries, complaints, and diseases is a skill that strongly determines the fate of our species. That is why people throughout all civilizations and ages have given much attention to the development of these skills. At the same time, in the past people struggled with an overwhelming shortage of knowledge that was far greater than the one that plagues people today. In an attempt to deal with their own ignorance they turned to religion and sorcery. Temperamental gods and witches casting all kinds of spells, alongside superstitions and taboos that go unrespected, represent just a fraction of the entire spectrum that encompasses unexplained causes of maladies. Mental disorders and psychosomatic diseases were particularly difficult to understand, and therefore more universal interpretations were ever more frequently formulated. Many of those early explanations were based on assumptions rather than scientific evidence. One of the most striking examples is the belief about the crucial role of the uterus in hysteria, one that was recognized in medicine as valid for 23 centuries. It is also common to find throughout history that the introduction of a new phenomenon has usually been followed by a rapid increase in the number of its cases being diagnosed. Patients usually start to "feel" or "observe" some of the symptoms, and thus doctors are enabled to "see" more of the disease. This may be explained by the fact that people have imagination and feelings. As a result, what we label as psychosomatic diseases may not really be accurate. For that reason, different diseases were prevalent in specific periods of Western medical history. Although some of the labels may appear naive, they all attempted to explain the mystery of psychosomatic medicine.

In the 18th century *vapors* were a concept that served to explain hysteric fits, and doctors believed that one third of all diseases were of nervous origin. The 19th century saw the term *nerves* indicated with increasing frequency as the cause of most psychosomatic problems. At the same time, hysteria became less popular as a diagnosis, giving way to neurasthenia, which was defined as a weakness of the nerves. The beginning of the 20th century was marked by the emergence of the concept of *nervous tension* that was soon replaced with the notion of stress. With this historical context in mind, stress, like all these previously mentioned concepts, may just be another explanation in a long line of pretenders to the throne in the kingdom of universal ailments. Furthermore, it may also serve some social purposes, which were brilliantly described by Martensen:

> Neurasthenia was one of those wonderful 19th-century diagnostic entities that promised something for almost everyone involved. A disease with loads of symptoms and little, well, finally no, organic pathology, it satisfied a number of the conditions any nosologic category must meet if it is to be broadly applied. During its heyday, which lasted from the 1870s to the turn of the century, the diagnosis of neurasthenia provided patients with a scientifically (and, hence, I would argue, socially) legitimate explanation of their inability to perform their expected roles.[18]

The theory of stress probably provides a similarly universal explanation nowadays. Is this, however, a well-grounded and accurate justification or merely a sham, a creation of cargo cult priests?

One of the most common convictions about our health is that stress weakens our immune system. The obvious conclusion is that the more stress we experience, the less resistant to diseases we are. This belief is so strongly established that many people take it for granted that stress can actually be the cause of cancer and that positive thinking can fight it. Unfortunately, research results show us a much less clear picture of the relationship between the amount of stress and the strength of our immune system. One review on human studies states: "Where perception of environmental stressors has been assessed using such indicators as daily moods, hassles, anxiety, social support or various coping measures, there has generally been a correlation between high perceived stress and depression of humoral and/or cell-mediated immunological measures. Often the most significant correlation has been with depression of natural killer (NK) cells."[59]

But it is not always so. A psychologist from Columbia University, Judith Rabkin, and her collaborators examined a group of 124 homosexual men who had tested positive for HIV, the immunity-weakening virus that can lead to AIDS. Those supposed to be suffering the most severe depression, emotional distress, or stressful life events showed no greater reduction in the number of helper T-cells and no more advanced symptoms of HIV infection than the others.[60]

Sometimes researchers observe entirely opposite effects, as was the case in one of Firdaus Dhabhar's experiments. He exposed stressed and non-stressed rats to an allergic condition and found that the stressed rats produced a better immune response than the non-stressed rats, and were found to have more white cells in their skin. In other words, stress seemed to have enhanced, not suppressed, the immune response.

A number of meta-analyses have shown that, contrary to popular belief, there is no relationship between emotions, stress, and cancer incidence.[61] Moreover, scientists have failed to establish a relationship between positive attitudes as well as the emotional states and the cancer survival rate.[62] Similar findings were reported in Australia in 708 women diagnosed with breast cancer. This study lasted for eight years and showed no links at all between negative emotions, depression, anxiety, fear, anger, or a pessimistic approach and the expected survival rate.[63]

Robert Sapolsky, a well-known researcher in the field of stress, summarizes his comprehensive research review on the relation between stress and cancer as follows: "So collectively, we have, with the exception of two studies concerning one type of cancer, no overall suggestion that stress increases the risk of cancer in humans."[64]

Research that involves cancer sufferers has also uncovered totally opposite effects. One such study included 37,562 nurses in the United States and lasted eight years (1992–2000). Based on findings from the study, it was concluded that the risk of breast cancer incidence is in fact lower by 17% among women who experience a relatively large amount of stress at work. Even more striking results were stated by Danish investigators. If truth be told, their research was based on a smaller sample (just under 7,000 women); however, the study lasted twice as long. It was demonstrated that in the group reporting a relatively high level of stress in the workplace, the risk of breast cancer was as much as 40% lower.[65]

These discrepancies are well explained by Booth and Pennebaker, who warn that:

> The immune system was not designed by an expert in psychometrics. Psychologists, when first learning about PNI, tend to think of the "immune system" as a coherent construct like "intelligence," "negative affectivity," or "need for achievement." Psychological measures of these constructs are generally created in a deductive manner. Each item that measures the construct is assumed to correlate – however modestly – with other items that also tap the overarching construct. The immune system, and by extension immune function, is not a deductively determined process. Some immune measures, for example, are highly cor-

> related and others are independent. The degree to which a person is resistant to say, a particular cold virus may be completely uncorrelated with their ability to ward off hepatitis. In other words, there is no representative measure of human immune behavior, partly because of the limited access to immune components in humans, but more especially because the notion of a representative measure of the immune system is essentially meaningless. ... The immune system undergoes continual change and immunity is not a unidimensional variable. The experience of a particular stressor associated with a decrease in the number of helper T cells lymphocytes in the blood has too often been interpreted as an example of "stress suppressing the immune system." This is a little like claiming that the quality of a symphony orchestra diminishes when the violas play more softly. We have to be careful not to over interpret observed immune changes as evidence of suppression or enhancement.[66]

Our immune system plays a major part in the development of our resistance to diseases and it does so by using many instruments in an array of different configurations. Therefore, it is not enough to analyze just the general influence of stress on the immune system to conclude unequivocally whether the dogma about the destructive effect of stress on our health is true. To do this, we need to have a closer look at particular health problems that are commonly perceived as stress-related conditions. Two of them are especially well-known and most people do not think that their causes could be anything other than stress. These are heart attack and peptic ulcers.

The first condition is frequently used as a literary trick by writers and screenwriters to mark the culmination of problems that affect the main character of a novel or a movie. Neither readers of the book nor spectators of the drama or movie have doubts as to the probability and veracity of the incident. This is hardly surprising since the heart has always been regarded as the seat of emotions. Do you recall that people did not object to the claim that the uterus was the source of hysteria for 23 centuries? I am afraid that there are similar misconceptions concerning the causes of myocardial infarction.

Even such a recognized researcher of stress as Robert Sapolsky tries to convince us that "The phenomenon is quite well documented," while in the next sentence he "proves" this in a completely incredible way: "In one study, a physician collected newspaper clippings on sudden cardiac death in 170 individuals."[67] Could there really be anything less credible than newspaper clippings? Is it possible to find a worse way of selecting cases for research? After all, what newspapers write about are oddities; the stranger the incident, the better suited it is for publication.

As a matter of fact, ischemic heart disease, which precedes myocardial infarction, is a disease of well-nourished people with a diet high in calories and saturated fat, the sort of diet prevalent in modern life. Moreover, it is usually combined with a low level of physical activity and sometimes with smoking. But, of course, that does not mean that stress itself cannot lead to myocardial infarction or cause sudden arrhythmic deaths. Research shows that most such incidents take place in the morning hours and often seem to come after heated arguments, exciting events, or strenuous exercise, hence the statement that they are triggered by stress may be true only if you analyze those cases cursorily. "However, when one looks more closely at the data relating to all of these events, one sees, for example, that the vast majority of the sudden deaths that occur to people after activities such as running for a bus, or pull-starting a lawn mower, and that those who die under these circumstances have engaged in these and similar activities many times before without dying. In addition, one sees that many of the people who die suddenly, die during or after activities such as urinating or defecating, which cannot easily be considered as "stresses" of modern life."[68]

Lawrence Hinkle, the author of the above citation, is not the only one who expressed some doubt regarding the role of stress as the cause of heart attacks. Donald and Ann Byrne, when looking at the effects of "stressful" life events, also found that the literature is inconsistent about whether life-change inventory scores predict any specific cardiovascular morbid events.[69]

David Goldstein from the National Institutes of Health has made a clear distinction between people with healthy and diseased hearts. He is confident that the relationship between chronic emotional stress and the development of hypertension, coronary disease, or any cardiovascular disorder in otherwise healthy people is unknown.[70]

All the available data suggest that heart disease starts to develop when men are in their teens and early twenties, particularly if they have a family history of the illness, drank alcohol, exercised little or ate badly. What is very seldom taken into account is the fact that, independently of the amount of stress in everyday life, heart disease in young women is extremely rare because their hormones protect them against it. A good recapitulation of the problem of stress and heart attacks was proposed by Serge Doublet:

> Again, there is no direct evidence of stress, in the guise of "modern life" or of its more specific "stressors", being the cause of heart disease. There is much anecdotal evidence, much speculation and assumptions but little proof. There is some evidence that people with an already diseased heart can suffer a heart attack. However, to suggest that in this situation, stress or any other factors, triggered the heart attack is like proposing that one drop of water (the last one) caused water in a glass to overflow, when it was the collective presence of many other drops that had made the occurrence of such an event possible in the first place.

The second disease which for decades was understood as a result of stress is the peptic ulcer. In all likelihood, the origins of this theory emanate from early research conducted in Selye's laboratory when, while conducting autopsies of dead animals used for experiments, he identified three common symptoms, with stomach ulcers being one of them. Although there had been some disagreement amongst scientists as to whether stress could really cause ulcers, the common belief as to the reliability of this claim was indisputable. However, one day in 1983 this changed due to the remarkable discovery of the bacterium *Helicobacter pylori* by two scientists, Barry Marshall and Robin Warren.[71] The microbe proved to be responsible for the development of peptic ulcers. Soon after, other researchers experimentally confirmed that *H. pylori* was present in nearly 100% of patients who have duodenal ulcers and 80% of those with gastric ulcers.[72] As a consequence, the U.S. Food and Drug Administration approved five *H. pylori* treatment regimens.[73] In 2005, the Nobel Prize Committee awarded Barry Marshall and Robin Warren the Nobel Prize in Physiology or Medicine for their discovery that *H. pylori* is the cause of peptic ulcers.

Nevertheless, official recognition seems to be insufficient to change the beliefs of the general public and, even worse, physicians about the causes of peptic ulcers. This was demonstrated by the results of U.S. national surveys of primary-care physicians and gastroenterologists on their knowledge of the association between *H. pylori* infection and peptic ulcers, carried out in 1994 and 1996. While as many as about 90% of physicians, when asked in the survey, identified *H. pylori* infection as the primary cause of peptic ulcers, they reported treating approximately 50% of patients suffering from first-time ulcer symptoms with anti-secretory agents without first testing for *H. pylori*. Gastroenterologists reported using treatment based on these agents in approximately only 30% of such patients.[74]

Another study performed in 1995, this time by the American Digestive Health Foundation and Opinion Research Corporation, revealed that 72% of the public were unaware of the association between *H. pylori* and ulcers.[75] Two years later, in 1997, a survey conducted by the U.S. Department of Health and Human Services revealed that approximately 60% of respondents still believed that ulcers were caused by stress, while another 17% thought that their main cause was the consumption of spicy foods. Only 27% were aware that a bacterial infection caused ulcers.[76]

At this point let me emphasize that the discovery of *H. pylori* has resulted in the eradication of ulcers in most of the cases when the correct treatment was applied, and the level of stress had no impact at all on the treatment and an eventual cure. It seemed that the discovery of the bacterium supported by the facts presented above would bring an end to research whose purpose was to prove the role of stress in the formation of ulcers. One must remember, however, that what is natural and well-understood among rational scientists is different among cargo cult priests. It is important to bear in mind that research on peptic ulcers had started in the mid-1950s, and although it can hardly be said to have been successful, many scientific articles have been published since then, while many people have made an academic career in the field of stress and peptic ulcers. A cargo cult does not cease to exist merely due to a lack of effect. Why would it have to stop?

When investigating articles in the EBSCO database, which mostly contains reports from the social sciences, from the period of 1983 (the publication date of the discovery of *H. pylori* as the cause of peptic ulcers) until December 2014, I found 228 peer-reviewed scientific articles devoted to the relationship between stress and ulcers. Many of these authors engage themselves in the process of demonstrating that there is solid evidence to show how psychological stress triggers many ulcers and impairs responses to treatment, while the presence of helicobacter is insufficient as a mono-causal explanation as most infected people do not develop ulcers. The authors maintain that stress probably functions as a cofactor with *H pylori*. And although little empirical evidence exists on the relation between stress and *H pylori*, this does not stop them from speculating about such a connection. The only pattern that has been found is a multitude of correlations. Oftentimes, cargo cult priests eager to find some indications that stress may play a role in the development of ulcers have elevated these correlations to the status of evidence.

One may wonder what their objective is. Even if stress were a contributing factor in the activation of *H. pylori*, this does not have any impact on the treatment. The only field where such a thesis might be useful is prevention, but prevention would require a lifetime of stress avoidance, something impossible to achieve as stress is inevitable. In this situation an antibiotic treatment might be far easier and much cheaper than prevention. However, there is one flaw in this approach – it means no work and no place for cargo cult priests, whose mission is to make people believe that stress is harmful and who earn money from helping patients avoid allegedly adverse impact of stress.

Summarizing his own review of the relationship between stress and health, Serge Doublet concluded: "There is no evidence to support the view that stress causes disease. To invoke the much weaker position that stress is a cofactor or a contributing factor may be an attempt to hide the lack of evidence by confounding stress with other factors."[77] Robert Sapolsky seems to share this opinion: "Everything bad in human health now is not caused by stress, nor is it in our power to cure ourselves of all our worst medical nightmares merely by reducing stress and thinking healthy thoughts full of courage and spirit and love. Would it were so. And shame on whose who would sell this view."[78] Lawrence Hinkle, a pioneering investigator of the mechanism of sudden death and of coronary heart disease, after reviewing the available evidence, concluded that, "So far as I am aware, the data that would allow one to obtain a quantitative answer to this hypothesis about the relation of 'stress' to illness in modern society do not exist."[79]

And yet, in what I have written so far lies the fundamental contradiction which needs to be revealed and explained. If stress does not influence our health or influences it to just a small extent, how is it possible that mere imagination can kill us, as was presented at the beginning of this chapter?

How does it come to pass that an imaginary fear can produce a stronger impact than the real fear that is an inherent component of many stressful situations? Is this not a paradox? To understand this, we need to take a more in-depth look at the phenomenon of voodoo death.

Although the concept of voodoo death has been present in literature for a long time, despite the popular belief in its effect it is not quite obvious whether it really exists. At the time when Walter Cannon introduced the term voodoo death, anthropologists were limited in their investigations of voodoo death by many factors. First, many of them had never witnessed any cases first hand and so had to rely on cases reported by others who usually had not witnessed any cases either. Almost none of those reports received medical examinations or autopsies. As Sapolsky writes: "Both Canon and Richter kept their theories unsullied by never examining anyone who had died of psychophysiological death, voodoo or otherwise."[80]

In 1961 Theodore Barber, a well-known expert in the field of hypnosis and suggestibility, criticized the concept of death by suggestion. In his opinion, in the reported cases the possibility of poison had rarely been ruled out based on toxicological examination. He also noticed that it might also have been the case that the person placing the curse on another furtively poisoned the victim (or killed the victim using some other method), thereby proving the sorcerer's tremendous power to all in the community. Moreover, in the majority of cases it appeared that the hexed individual had refused food and water and, in other cases, family members had refused to give the victim food and water. All those factors could have contributed significantly to the person's eventual death. Some instances of voodoo death were also probably due to organic illness.[81]

Moreover, it was not only how the reported cases were interpreted that was criticized. David Lester calls into question the value of empirical research in this field: "In Cannon and Richter's research, which was on animals and not on humans, the death occurs suddenly. Voodoo death does not occur suddenly, but rather takes several days to occur. As was noted above, Cannon's animals died after intense hyperactivity which, again, is not characteristic of voodoo death. Richter's rats had their whiskers trimmed (a trauma) and made to swim. They died quickly in this stressful situation, which is not characteristic of voodoo death."[82]

It is highly probable that if so-called voodoo death is a fact-based phenomenon, it should be ascribed to reasons different than the effect of suggestion. Among many possible explanations for this type of death, the abandonment of actions necessary to sustain life, such as drinking and eating, is one of the most probable.

But how can these alternative justifications of voodoo death help us in understanding instances of premature death of people who believe that stress influences their health? This is how the authors of the previously cited research see this problem:

An individual's health locus of control, defined as their beliefs in the control they have over their own health (Wallston, Stein, & Smith, 1994), may also contribute to a heightened perception of the health implications of stress. Those who perceive that stress affects their health may have an external locus of control, believing that their health is not in their control, but attributable to external circumstances. Studies have indicated that individuals who have a high external locus of control experience worse outcomes than those who feel that their health is within their control (Heath, Saliba, Mahmassani, Major, & Khoury, 2008; Preau et al., 2005). Although much of this research has focused on those with an illness, the present study suggests that health-related locus of control (as seen in a greater perceived impact of stress on health) may also contribute to outcomes in healthy populations. As such, encouraging active attempts at problem solving and increasing an individual's sense of control over their stress levels and health may potentially lead to better health outcomes by allowing individuals to better utilize coping resources (Thoits, 1995).[83]

Here we are, then. Hexed individuals may act in a way that prompts their own death. The most obvious possibility is dehydration. Individuals in the research carried out by Abiola Keller and her collaborators were also hexed by the unceasing propaganda of cargo cult priests. The very propaganda whose main claim is that stress has a negative impact on our health. Similarly to the natives from New Zealand, the victims of this propaganda do not believe that what happens with their health is under their control. Consequently, they do not care about the quality of their diet very much and do not appreciate the benefits of physical activity. To all appearances, they undervalue the negative impact of smoking and drinking, and they do not follow the recommendations of their doctors. And they die easier and faster than those who are not hexed.

In this way, cargo cult priests are able to elicit the necessary helplessness referred to by John Holt, whom I quoted in the previous chapter. The concept of stress represents a very specific instance of science being changed into what Feynman termed cargo cult science. Still, having only this one example it would be difficult to draw inferences about the general causes responsible for such transformations observed in the entire domain of psychology. In order to find other contributing factors, the next chapter looks at our discipline from a broader perspective so we can examine what represents the actual subject of its study.

---

[44] Tabooed place – my note.
[45] God – my note.
[46] W. Brown, *New Zealand and Its Aborigines: Being an Account of the Aborigines, Trade, and Resources of the Colony; And the Advantages it now Presents as a Field for Emigration and the Investments of Capital* (London: Smith, Elder, and Co., 1845): 76-77.
[47] W. B. Cannon, "'Voodoo' Death," *American Anthropologist 44*, (1942): 169–181, 169.

[48] A. Keller, K. Litzelman, L. E. Wisk, T. Maddox, E. R. Cheng, P. D. Creswell, W. P. Witt, "Does the Perception that Stress Affects Health Matter? The Association with Health and Mortality," *Health Psychology 31*, (2012): 677–684.
[49] H. Nabi, M. Kivimäki, G. D. Batty, M. J. Shipley, A. Britton, E. J. Brunner, J. Vahtera, C. Lemogne, Alexis Elbaz, and A. Singh-Manoux, "Increased Risk of Coronary Heart Disease Among Individuals Reporting Adverse Impact of Stress on their Health: the Whitehall II Prospective Cohort Study," *European Heart Journal 34*, (2013): 2697–2705.
[50] S. Dooblet, *The Stress Myth*. (Chesterfield MO: Science and Humanities, 1999).
[51] Selye used to claim that he was the first to use the word *stress* in a biomedical, rather than an engineering sense. Actually, Walter Cannon used this term decades earlier; W. Cannon, "The Interrelations of Emotions as Suggested by Recent Physiological Researches," *American Journal of Psychology 25*, (1914): 256.
[52] H. Selye, *The Stress of Life* (New York: McGraw-Hill, 1956).
[53] Dooblet, *The Stress Myth*.
[54] Cannon, "The Interrelations of Emotions."
[55] R. M. Sapolsky, *Why Zebras Don't Get Ulcers* (New York: Henry Holt, 2004), see chap. 1, note vii.
[56] Dooblet, *The Stress Myth*.
[57] For detailed review see, ibid.
[58] R. L. Martensen, "Was Neurasthenia a 'Legitimate Morbid Entity?'" *The Journal of the American Medical Association 271*, (1994): 243.
[59] J. R. Booth, "Stress and the Immune System," in *The Encyclopaedia of Immunology*. 2nd ed. ed. I. M. Roitt, and P. J. Delves, (London: Academic Press, 1998), 2220-2228.
[60] J. G. Rabkin, J. B. W. Williams, RH, Remien et al. "Depression, Distress, lymphocyte Subsets and HIV Symptoms on two Occasions in HIV-positive Homosexual Men,"*Archives of General Psychiatry 48*, (1991): 111-119.
[61] P. N. Butow, J. E. Hiller, M. A. Price, S. V. Thackway, A. Kricker, and C. C. Tennant, "Epidemiological Evidence for a Relationship Between Life Events, Coping Style, and Personality Factors in the Development of Breast Cancer," *Journal of Psychosomatic Research 49*, (2000): 169-181; S. F. A. Duijts, M. P. A. Zeegers, and B. V. Borne, "The Association Between Stressful Life Events and Breast Cancer Risk: A Meta-analysis," *International Journal of Cancer 107*, (2003): 1023-1029; M. Petticrew, J. M. Fraser, and M. F. Regan, "Adverse Life-events and Risk of Breast Cancer: A Meta-analysis," *British Journal of Health Psychology 4*, (1999): 1-17.
[62] B. L. Beyerstein, W. I. Sampson, Z. Stojanovic, and J. Handel, "Can Mind Conquer Cancer?" in *Tall Tales About the Mind and Brain: Separating Fact from Fiction*, ed. S. Della Sala (Oxford: Oxford University Press, 2007), 440-460.
[63] K. A. Philips, "Psychosocial Factors and Survival of Young Women with Breast Cancer," Annual Meeting of the American Society of Clinical Oncology, Chicago, (June 2008).
[64] Sapolsky, *Why Zebras*, 148.
[65] N. R. Nielsen, Z. F. Zhang, T. S. Kristensen, B. Netterstrom, P. Schnor, and M. Gronbaek, "Self-reported Stress and Risk of Breast Cancer: Prospective Cohort Study," *British Medical Journal 331*, (2005): 548.

[66] R. J. Booth, and J. W. Pennebaker "Emotions and Immunity," in *Handbook of Emotions. 2nd Ed.*, ed. M. Lewis, and J. M. Haviland-Jones (New York: The Guilford Press, 2000), 558-572, 560.
[67] Sapolsky *Why Zebras*, 49.
[68] L. E. Hinkle Jr., "Stress and disease: The concept after 50 years," *Social Science and Medicine 25*, (1987): 561-566.
[69] D. G. Byrne, and A. E. Byrne, "Anxiety and Coronary Heart Disease," in *Anxiety and the Heart*, ed D. G. Byrne and H. Rosenman, (New York: Hemisphere, 1990), 213-232.
[70] D. S. Goldstein, *Stress, Catecholamines, and Cardiovascular Disease* (New York: Oxford University Press, 1995).
[71] J. Warren, and B. Marshall, "Unidentified Curved Bacilli on Gastric Epithelium in Active Chronic Gastritis," *The Lancet 1*, (1983): 1273.
[72] A. Ateshkadi, N. Lam, and C. Johnson, "*Helicobacter pylori* in Peptic Ulcer Disease," *Clinical. Pharmacy 12*, (1993): 34.
[73] R. J. Hopkins, "Current FDA-approved Treatments for Helicobacter Pylori and the FDA Approval Process," *Gastroenterology 113*, (1997): 126-30.
[74] P. Novelli, "Knowledge About Causes of Peptic Ulcer Disease -- United States, March-April 1997," *Weekly*, *46*, (October 1997): 985-987, http://www.cdc.gov/mmwr/preview/mmwrhtml/00049679.htm
[75] American Digestive Health Foundation and Opinion Research Corporation, "Familiarity with *H. Pylori* Among Adults with Digestive Disorders and their Views Toward Diagnostic and Treatment Options," (Bethesda, MD: American Digestive Health Foundation and Opinion Research Corporation, 1995).
[76] Novelli, "Knowledge About Causes."
[77] Dooblet, *The Stress Myth*, 225.
[78] Sapolsky, *Why Zebras*, 154.
[79] Hinkle, "Stress and Disease," 566.
[80] Sapolsky, *Why Zebras*, 55.
[81] T. X. Barber, "Death by Suggestion," *Psychosomatic Medicine 23*, (1961): 153-155.
[82] D. Lester, "Voodoo Death," *Omega, 59*, (2009): 1-18.
[83] Keller, "Does the Perception."

Chapter 3: Self Reports and Finger Movements of the Weirdest People in the World

Cargo cults arose from the misery of life their future followers had been leading at a time when more fortunate visitors, bathed in the luxury of all kinds of wealth, set foot on their land. The gap between the two groups was so remarkable that it could only have been ascribed to supernatural forces. The New Guineans chose the only interpretation of reality that they could comprehend.

Psychology came into being in an era of resounding triumphs of science and technology. The establishment of the world's first psychology laboratory, which was founded at the University of Leipzig by German psychologist Wilhelm Wundt in 1879, is considered to mark the emergence of psychology as an independent empirical discipline. Rail transport had already conquered the world by then and the first cars were being produced. The theory of evolution had been known for more than twenty years, Mendeleev's periodic table of elements had been created ten years earlier, and the telephone was two years old. Early in the year when Wundt created the first psychological lab, Thomas Alva Edison filed his patent for a lightbulb, and already in December he presented to the public his system of electric illumination. A year later Alexander Graham Bell used his photophone to make the first ever wireless transmission of sound. In 1882, Robert Koch discovered the bacteria that causes tuberculosis; in 1885, Louis Pasteur vaccinated the first patients against rabies; and in, 1887 Herman Hollerith obtained a patent for his tabulator based on punched cards as data carriers. At that time psychology barely had just developed a process for systematization of a method for recording subjective thoughts and feelings, knowns as introspection.

Towards the end of the 19th century William James, a philosopher and psychologist who would come to be known as "the father of American Psychology," lamented the status of psychology as a field of science: "A string of raw facts; a little gossip and wrangle about opinions; a little classification and generalization on the mere descriptive level; a strong prejudice that we have states of mind, and that our brain conditions them: but not a single law in the sense in which physics shows us laws, not a single proposition from which any consequence can causally he deduced. This is no science, it is only the hope of a science."[84]

This contrast between empirical science, which was enjoying an abundance of scientific breakthroughs and technological inventions, and psychology, which had only just been born and could only point to the scarcity of its own accomplishments, is deceptively similar to the situation experienced by the poor islanders and the wealthy strangers. It is nothing but heroic attempts to rival glorious achievements of science that led psychology astray. As sad as it is, almost none of those responsible for the development of these subjects of study that were from the outset doomed to failure was able to bring themselves to admit that they had wasted time and energy looking for answers where they by no means could have been found. This is how the first cargo cult sciences came to life, such as psychoanalysis. Their leaders remain trapped to this day, marking their time by performing rituals whose outcome is as inevitably evident as the outcome of the efforts made by cargo cult followers.

However, the fallacies of the originators of psychological fields of study are not that intriguing. At the end of the day, had it not been for them, there would have been no progress at all. It is only the subsequent sluggishness that spread, despite important discoveries which could have propelled our discipline forward, which seems to be a puzzle worth solving.

Unfortunately, a poverty of accomplishments in psychology as compared to other, even related sciences, is not only distant history. Quite recently, in 1990 "the Decade of the Brain" was announced by U.S. President George H. W. Bush as part of a larger effort involving the Library of Congress and the National Institute of Mental Health of the National Institutes of Health "to enhance public awareness of the benefits to be derived from brain research."[85]

The efforts put into understanding the workings of the human brain were very fruitful. This is how the editors of *Acta Psychologica* summarize this decade: "The amount of knowledge gained from investigating the neural basis of behavior and cognition was unprecedented. As a result, a new interdisciplinary field emerged, the cognitive neurosciences, specifically devoted to understanding brain-mind relationships. Tremendous progress has been made: five years after the first monumental edition of *The Cognitive Neurosciences* (Gazzaniga, 1995, MIT Press), a second equally monumental (and almost completely new) edition was published, titled *The New Cognitive Neurosciences* (Gazzaniga, 2000, MIT Press)."[86]

This is not, however, the whole story. As the authors put it in a subsequent part of their article:

> Some fainthearted will be daunted by so much newly accumulated knowledge in such a short time span and would want to give up attempting to follow that progress. But then others would happily take over. In fact, some neuroscientists have already been so enthusiastic about this progress that they forget that there is more to understanding behavior and cognition than tracing the neural activities alone. For example, in his *Principle of Neural Science* the most recent Nobel laureate in medicine, Erick Kandel, states that "the goal of neural science is to understand the mind – how we perceive, move, think, and remember" (Kandel, Schwartz, & Jessell, 2000, p. xxxv). Some psychologists will be shocked by this kind of statement because understanding the mind was supposed to be the traditional territory of psychologists.
> In reaction perhaps, the American Psychological Association declared the new decade "the decade of behavior."[87]

My intention is not to decide whether the establishment of the decade of behavior was the result of a shock caused by the accomplishments of the decade of the brain. Personally, I am far less enthusiastic about the advancements of neuroscience than the editors of *Acta Psychologica*. This discipline sometimes assumes the form of a cargo cult as well, which I will explore in Chapter 5.

Nevertheless, the decade of behavior came immediately after the decade of the brain and was announced in grand style:

> At press time, a bipartisan group of 10 senators had signed a letter asking President Clinton to make it official and name 2000–2010 the Decade of Behavior. But even without the president's endorsement, the Decade will begin, says Richard McCarty, PhD, executive director of APA's Science Directorate. The National Advisory Committee for the Decade of Behavior, which includes representatives from psychology, sociology, geography, anthropology, political science, economics and public health, has gained support for the initiative from 17 federal funding agencies.[88]

To my knowledge, President Clinton never made a formal endorsement of the decade, but it is certainly worth finding out what psychologists have learned and gained from these ten years of research.

Only in the middle of this decade did Roy Baumeister and his co-workers decide to verify whether psychologists are really capable of examining behavior. To do this, they analyzed the content of the January 2006 issue of the *Journal of Personality and Social Psychology* (JPSP), one of the most prominent psychological journals. The issue contained 11 articles reporting 38 studies. Apart from one case, not a single one of those 38 studies contained direct observation of behavior. The dependent measures consisted entirely of introspective self-report ratings, either on paper or computer-administered questionnaires.[89]

These findings are all the more striking when we consider that for decades now psychology students have been taught that psychology is the science of behavior and that its primary goal is to describe and explain human behavior. The same description of our discipline has been given in most psychology textbooks, with some of them having even appropriately corresponding titles, such as *Psychology: the Science of Behavior*.

To make sure that their discovery was not mere chance, Baumeister and his co-workers decided to examine another previous issue of the JPSP. Again, there were only two out of 38 studies in which behavior was observed. To form a more systematic and quantified view of this problem, the investigators decided to analyze the content of JPSP over decades. To this end, they selected March and May issues from 1966, 1976, 1986, 1996 and 2006. This is what they found: "Back in 1966, when most articles contained only a single study, about half of these involved actual behavior. The study of behavior increased its share of the journal into the 1970s. But the use and study of behavior dropped sharply in 1986, and the subsequent decades have seen a continued downward trend. Apparently, the study of behavior has been in a steady decline since the early 1980s."[90]

It seems that the decade of behavior did not change much in this respect, as in 2009 one of the best-known field researchers in social psychology, Robert Cialdini, published an article that can be seen as a kind of desperate confession. In the first sentence of his paper he announced: "I am planning to retire early from my university psychology department position."[91] He gave three reasons for his decision: "(a) the advent of the cognitive revolution, (b) the unwritten (but nearly iron-clad) requirement for multiple study packages in our very top journals, and (c) the prioritization of mediational analysis of one's effects through the use of secondary measures."[92] Although he admitted there was considerable value in each of these trends, he also noted that his field was experiments in naturally occurring settings with behavior as the prime dependent variable. However, that experimental orientation does not fit the developments he listed.

History has now come full circle and we are again in the reality created by JPSP. Cialdini wrote:

> The flagship journal in social psychology is the Journal of Personality and Social Psychology (JPSP). As a past Associate Editor, I know how to get papers accepted there; along with my coworkers, I've continued to have JPSP articles published regularly in recent years. But, I haven't had any of my field research published there in over 15 years. And field research, remember, is what I do best. So, I have had to take reports of that work elsewhere, sometimes to top-of-the-line scholarly outlets in consumer behavior, organizational science, survey research, marketing, and management.[93]

Was Cialdini offended by the changes that had occurred in psychology? Not necessarily. He points to serious consequences of this shift which, in his opinion, have been devastating at least in one respect: "I am no longer able to accept graduate students. At least, I am no longer able to do so in good faith because most apply hoping (a) to be trained by me in field research methods for investigating behavior in naturally occurring settings and (b) to be competitive for the best jobs in academic social psychology at the end of that training. For the foreseeable future, I know that I can reasonably help them attain only the first of those goals."[94]

Cialdini's confessions revealed not only his personal dilemmas. What he unearthed was, above all, the most critical problems plaguing psychology today. The failure of the profession to address and solve them adds to the transformation of psychology into a cargo cult. This was well shown by West and Brown, who in 1975 conducted the same experiment twice, in two different ways. In the first trial they asked a group of individuals to indicate what their behavior would be in a specific situation. In the second study, they actually staged the whole event. The experiment involved an alleged accident victim asking people on the street for money to help pay for medical care at a nearby clinic. Perhaps unsurprisingly, the levels of donations were dramatically different; the declarations were far more generous than actual donations given to the "victim." Moreover, the victim's attractiveness did not have a significant effect on hypothetical donations, but it did have a significant effect on real donations. The results of this research unambiguously show how worthless conclusions about human behavior are if they are drawn only from declarations.

Unfortunately, people's predictions about how they will react and feel[95] or how they will make decisions[96] are similarly worthless. They are not far from the conclusions formulated by cargo cult believers. But this was not always so. Baumeister and his associated state:

> The move from behavior to an emphasis on introspective self-report and hypothetical responses to imagined events is potentially a hugely important shift in the very nature of psychology. Psychological science started out in the 1800s with introspection (e.g., Wundt, 1894). One major development of the 20th century was the shift from introspection to direct observation of behavior, widely regarded as an advance in the development of scientific methodology. Did someone, somewhere, decide that that had been a mistake and that we should now go back to introspection?[97]

Unfortunately, it is the scientists themselves who decided to undertake this shift. It is far easier to make an academic career examining self-reports than it is studying behavior. It takes much less energy to carry out research in a laboratory than in natural settings. It is much cheaper to ask for self-reports than to stage experiments designed to observe behavior. And all this, instead of developing our discipline into a domain of empirical knowledge, is little by little changing it into a cargo cult.

However, it is a curious fact that if someone wished to bury themselves in an exhaustive review of the efforts of thousands of scientists working on behavior throughout the decade of behavior, they will find nothing but emptiness. No enthusiastic speeches, no reports enumerating endless achievements, no conferences or banquets celebrating accomplishments. The website launched by The National Advisory Committee for the Decade of Behavior (www.decadeofbehavior.org) is empty now. The links to pages created by the APA with announcements of this initiative have vanished into thin air. In my own pursuit of any summary of achievements I searched the EBSCO database for scientific articles published in the years 2010–2015 that contained the phrase "decade of behavior" or "decade of behaviour." What I found was only 11 papers matching my search criteria. Rather than lists of scientific accomplishments, the results comprised just a few minor mentions in editorials and articles. Are some ashamed of how this decade has ended?

All the same, there is one element of the Decade of Behavior that is worth mentioning. This is the Books in Decade of Behavior Series created by the APA, and it will certainly last for a long time.[98] However, the criteria for selecting books seem somewhat incomprehensible. The series offers many interesting titles with regard to visual perception, intelligence, cognitive processes, self-interest, social categories, memory, gender differences, and other academic areas of interest. Are all these fields of research called behavior now?

However, as Baumeister and his co-workers have shown, behavior (as understood in old-fashion terms, not as described above) is so rare a subject of psychological research that it is not a coincidence that most examples of reliable psychological practices which we can identify (the ones that will be addressed in the last part of this book, in the chapter entitled "A Letter to Richard Feynman") are taken from the disciplines where behavior is analyzed. In many areas of our functioning neither behavior, introspection, nor cognitive mediating variables are of crucial importance. This is the case in health psychology. "It has been argued that 48% of deaths in the USA are attributable to behavior, the 'all consequence' behavioral risk factors being smoking, physical activity, dietary behaviors and alcohol intake."[99] Even in preventing cancers, behavioral factors have been evaluated as constituting as much as 70% of all causes.[100] Despite the fact that behavior is of such significance from the point of view of health psychology, researchers focus on intra-psychic phenomena, such as thoughts and emotions, rather than behavior.[101] To make a psychological theory useful for implementing evidence-based practices in healthcare, researchers must overcome the many obstacles that pile up. A consensus group composed of healthcare professionals within a project implementing psychological theory into behavior change practices recognized 33 psychological theories explaining behavior with 128 theoretical explanatory constructs.[102] Unfortunately, this list was not the result of an exhaustive literature review, but merely the outcome of brainstorming sessions held by this consensus group. It is highly unlikely that we need so many theories to explain behavior. This manifestation of the proliferative character of our discipline results in overlapping theories. Its constructs are most often variants of those that have already been developed. It seems that an enormous amount of energy has been used in trying to reinvent the wheel.

When I think about this problem, I imagine a garage cluttered up with many unnecessary tools. The space is full of used parts of unknown machines and other accessories everyone thinks are useful but no one knows what they are really designed for. Several centuries ago the English philosopher William of Ockham proposed a very useful scientific tool, nowadays known as "Ockham's razor." We can use it to introduce order into our discipline, but it seems that almost nobody is truly concerned to do so. What we really need in psychology nowadays is a minimalistic approach. However, it is beyond the capacities of the cargo cult believers. All these countless theories and constructs are very useful to develop and fuel our next rituals. Looking for correlations between them, explaining interactions, presenting them at conferences, earning scientific titles – all these activities are the building blocks of a cargo cult system. To take these toys away from cargo cult followers is to take the meaning of life away from them.

Regrettably, the misery is not simply attributable to the lack of behavior in the science of behavior and overlapping scientific theories. In 2010, another painful blow to the arrogant psychology of the Western world was the publication of a famous article titled "The Weirdest People in the World."[103] The article was written by three scientists from the University of British Columbia, Joseph Heinrich, Steven J. Heine, and Ara Norenzayan. They demonstrated that the output of contemporary psychology may be relevant only to an exotic group of societies which they called WEIRD. This catchy name is in fact a play on words since the adjective *weird* is at the same time an acronym that consists of the first letters of the adjectives *Western*, *Educated*, *Industrialised*, *Rich*, and *Democratic*.

The analysis of research studies in six sub-disciplines of psychology published in the top journals from 2003 to 2007 carried out by Heinrich and his colleges shows that the 96% of the subjects came from industrialized Western countries which make up just 12% of the world's population. Of this number, up to 68% were from the United States of America. Interestingly, the authors also examined the content of the premier journal in social psychology – the JPSP. They revealed that 67% of the American samples (and 80% of the samples from other countries) were composed solely of undergraduates in psychology courses, which means that the chance that the research subject was a randomly chosen American student is 4,000 times greater than it was someone from beyond the circle of Western civilization. As a result, our knowledge about human nature is based on a little group that the researchers have given the name WEIRD.

However, the lack of representativeness in psychological research is not the main cause for concern. After all, it may turn out that in many aspects human nature is consistent, and that the way we see, for example, proportions does not depend on whether we live in the United States or on the Kirghiz steppe. Unfortunately, this is not the case. The authors have also conducted a comparative analysis of the obtained results in many different fields, such as visual perception, fairness, cooperation, spatial reasoning, categorization and inferential induction, moral reasoning, reasoning styles, self-concepts and related motivations, and the heritability of intelligence. It turned out that the people referred to as WEIRD strongly differ from other populations, and they also often obtain extreme scores. The most surprising thing is that such apparently basic cognitive processes as size or shape perception are different in many cultures.

A striking example of this is the study of the Müller-Lyer illusion illustrated in Figure 1. Since the author first described it in 1889, it has no doubt found its place in every psychology textbook, at least in those dealing with perceptual regularities. Most of us see the upper line as longer than the lower one, whereas in fact they are identical. But some studies, which show that it is not necessarily an illusion, and if it is, it is only for us – people from WEIRD societies – have the effect of pouring a bucket of cold water on someone's head. Members of the San tribe from the Kalahari valley see these lines as equal and are completely undeceived. When we ask people from different cultures about the differences in the lengths of the lines, we find the highest scores (the largest discrepancy) when studying American students. Could it be that, as suggested by Heinrich and his colleagues, the alleged illusion was only a by-product of our culture?

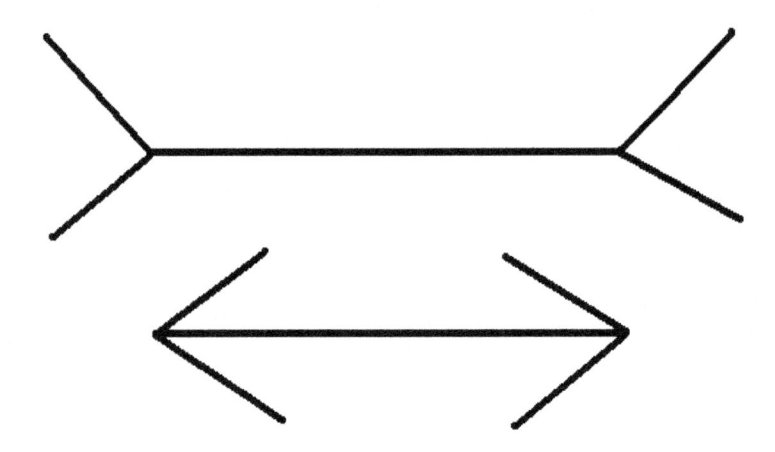

This example comes from groups of comparisons made by the authors between industrialized societies and small-scale societies. Within the first group, called "Contrast 1," they also pointed to huge differences in fairness and cooperation in economic decision-making, folk biological categorization, reasoning, and induction, as well as in spatial cognition. In the summary of this part of their analysis they concluded: "Given all this, it seems problematic to generalize from industrialized populations to humans more broadly, in the absence of supportive empirical evidence."[104]

The second part of their analysis, titled "Contrast 2," was devoted to comparisons between Western and non-Western samples. The authors examined four of the most studied domains: social decision-making (fairness, cooperation, and punishment), independent versus interdependent self-concepts (and associated motivations), analytic versus holistic reasoning, and moral reasoning. Although Heinrich and his colleagues noticed that robust patterns have emerged among people from industrialized societies, Westerners are shown to be unusual and frequent global outliers in several key dimensions. Many of these differences were not merely differences in the magnitude of effects but often showed qualitative differences, involving effect reversals or novel phenomena such as allocentric spatial reasoning and antisocial punishment.

Given the dominance of American research within psychology and the behavioral sciences, the authors decided that it would also be important to assess the similarity of American data with that from Westerners more generally. The part of the article titled "Contrast 3: Contemporary Americans Versus the Rest of the West" revealed that American participants are exceptional even within the unusual population of Westerners; they were, as they called them, "outliers among outliers."[105]

The majority of behavioral research on non-clinical populations within North America is conducted with undergraduates. Therefore "Contrast 4" investigated how typical contemporary American subjects compared with other Americans. This is what it revealed:

> Some documented changes among Americans over the past few decades include increasing individualism, as indicated by increasingly solitary lifestyles dominated by individual-centered activities and a decrease in group participation (Putnam 2000), increasingly positive self-esteem (Twenge & Campbell 2001), and a lower need for social approval (Twenge & Im 2007). These findings suggest that the unusual nature of Americans in these domains, as we reviewed earlier, may be a relatively recent phenomenon. For example, Rozin (2003) found that attitudes towards tradition are more similar between Indian college students and American grandparents than they are between Indian and American college students. Although more research is needed to reach firm conclusions, these initial findings raise doubts as to whether research on contemporary American students (and WEIRD people more generally) is even extendable to American students of previous decades.[106]

The analysis performed by Heinrich and his colleagues exposed the fragility of the entire construct that we call psychology. One must keep in mind that the authors have analyzed only its fragments. In search of the truth about man and human nature, we will be struggling through natural barriers that arise in the course of research but also through heaps of papers with overgeneralized conclusions formulated by careless scientists. While overcoming such obstacles, we need to keep in mind what the authors of the analysis have highlighted in their closing words:

> The sample of contemporary Western undergraduates that so overwhelms our database is not just an extraordinarily restricted sample of humanity; it is frequently a distinct outlier vis-à-vis other global samples. It may represent the worst population on which to base our understanding of *Homo sapiens*. Behavioral scientists now face a choice – they can either acknowledge that their findings in many domains cannot be generalized beyond this unusual subpopulation (and leave it at that), or they can begin to take the difficult steps to building a broader, richer, and better-grounded understanding of our species.[107]

Several years ago I was giving lectures entitled "Psychology as a Cargo Cult Science." I usually began by showing a clip from a documentary about cargo cult believers. I played this clip without a word of introduction and observed the audience. For many people it was shocking to watch how cargo cult believers had built their runways in the middle of the jungle, how they had lit bonfires on them, and how serious they had been about those practices. The audience watched the movie in silence and in full concentration. However, as the lecture went on, most of them were not even partially shocked when they learned how scientists, rather than focusing on examining behavior, keep busy by asking us how we will behave in a number of situations and how, with the solemnity of the cargo cult believers, they calculate the results of such research, seek correlations, and diligently publish articles on this subject. As they do so, their blood boils whenever another self-report expert challenges their interpretation at some conference devoted almost entirely to self-assessment declarations of the analyzed subjects. In the same way, my audience was not alarmed when I commented further to explain all those declarations and introspective self-reports had been collected from members of a small tribe that is called WEIRD because it differs so much from the rest of the population. And even if a group of scientists eventually turns their attention to examining actual behavior, it is hardly possible to discern what is valuable and useful within the myriad of theories and constructs they create. I am painfully aware that the majority of the listeners did not see the analogy between the practices of cargo cult believers depicted by me earlier and the scientists. They could perceive neither the pointless bustle of both the former and the latter, nor how fruitless their efforts were. Nevertheless, I invariably see the analogy in the behavior demonstrated by partakers in rituals that are performed both by the islanders and scientists. As a rule, the cargo priests have such a profound faith in the meaning of their efforts that they pass on this belief to those around them, irrespective of whether they wear loincloths and have their noses pierced or if they wear a white laboratory coat and put an academic degree before their name.

---

[84] W. James, *Psychology: Briefer Course* (New York: Holt, 1892), 468.
[85] E. G. Jones, and L. M. Mendell, "Assessing the Decade of the Brain," *Science 284* (April 1999): 739.
[86] Editorial, "Beyond the Decade of the Brain: Towards a Functional Neuroanatomy of the Mind," *Acta Psychologica 107*, (2001): 1-7.
[87] Ibid.
[88] B. Azar, "APA Ushers in the 'Decade of Behavior,'" *Monitor on Psychology 31*, (January 2000): http://www.apa.org/monitor/jan00/decade-of-behavior.aspx
[89] R. F. Baumeister, K. D. Vohs, and D. C. Funder, "Psychology as the Science of Self reports and Finger Movements," *Perspectives on Psychological Science 2*, (2007): 396-403.
[90] Ibid.
[91] R. B. Cialdini, "We Have to Break Up," *Perspectives on Psychological Science 4*, (2009): 5-6.
[92] Ibid.

[93] Ibid.
[94] Ibid.
[95] T. D. Wilson, and D. T. Gilbert, "Affective forecasting," in *Advances in Experimental Social Psychology 35*, ed. M. Zanna (New York: Elsevier, 2003): 345–411.
[96] C. A. Holt, and S. Laury, "Risk Aversion and Incentive Effects," *Andrew Young School of Policy Studies Research Paper Series 06-12*, (2002): ttp://ssrn.com/abstract=893797
[97] Baumeister, et. al., "Psychology as the Science," 397.
[98] http://www.apa.org/pubs/books/browse.aspx?query=&fq=DocumentType:%22Book/Monograph%22%20AND%20SeriesFilt:%22Decade%20of%20Behavior%20Series%22
[99] M. Johnston, and D. Dixon, "Current Issues and New Directions in Psychology and Health: What Happened to Behaviour in the Decade of Behaviour?" *Psychology and Health 23*, (2008): 509-513, 509
[100] R. Doll, and R. Peto, "The Causes of Cancer – Quantitative Estimates of Avoidable Risks of Cancer in the United-States Today," *Journal of the National Cancer Institute 66*, (1981): 1191–1308.
[101] J. Ogden, "Changing the Subject of Health Psychology." *Psychology and Health 10*, (1995): 257–265.
[102] S. Michie, M. Johnston, C. Abraham, R. Lawton, D.Parker, and A. Walker, "Making Psychological Theory Useful for Implementing Evidence Based Practice: A Consensus Approach," *Quality and Safety in Health Care 14*, (2005): 26–33.
[103] J. Heinrich, S. J. Heine and A. Norenzayan, "The Weirdest People in the World," *Behavioral and Brain Sciences 33*, (2010): 61-135.
[104] Ibid., 69.
[105] Ibid., 76.
[106] Ibid., 77-78.
[107] Ibid., 123.

CHAPTER 4: IGNORANCE, NONCHALANCE, OR CONSPIRACY? ACTUAL STATISTICAL SIGNIFICANCE OF RESEARCH RESULTS

Our statistics should not become substitutes instead of aids to thought.
—After Bakan

"In Hinton, West Virginia, one woman was killed by three bullets and her body dumped beside a remote road; another woman was beaten to death, run over by a car, and left in the gutter. Both had told people that they had AIDS, and the authorities said that this was the reason they had been killed."[108]

If a patient who tested HIV positive in the early 1980s did not meet a fate similar to the one experienced by the aforementioned women, they most often had to face losing their job, in many cases losing their family, and most definitely discrimination, severe stigmatization, and ostracism. For this reason, it is not surprising that when 22 blood donors in Florida were notified that they had tested HIV positive, seven of them committed suicide.[109] Such dramatic decisions were not isolated, inasmuch as the diagnostic tests that were in use had very high sensitivity and specificity (the parameters of which will be explained later in this chapter). "In Ohio, a man with positive HIV test lost, within 12 days, his job, his home, and-almost-his wife. The day he was going to commit suicide he was notified that he had received a false positive test result."[110] Was it a miraculous twist of fate that held back the touch of death at the very last moment? Not really, rather a typical case. A medical text that documented the tragedy of the blood donors from Florida many years later called the reader's attention to the fact that "even if the results of both AIDS tests, the ELISA and WB [Western blot], are positive, the chances are only 50-50 that the individual is infected."[111] What is of significance is that the Florida donors had been tested only with ELISA, which has a higher false positive rate than when combined with the Western blot. Allowing for this, as well as for the fact that they were people with low-risk behavior, the likelihood of them being infected was probably even lower than 50%. If the donors had been aware of the actual degree of risk of contracting HIV, they might still be alive today.[112]

Why have I chosen HIV diagnosis-related dilemmas as a discussion point in a psychology book? Well, because what propelled the donors into suicide was not the deadly virus itself but their ignorance of mathematics. What is even worse, it was also the ignorance of those physicians who had informed the donors about their test results that could be a contributing factor behind their decisions. Unfortunately, similar disregard of theoretical and methodological fundamentals is also not rare among scientists who engage in psychological experiments, study behavior, and diagnose people.

The HIV diagnosis case should be striking enough to turn our attention to the problem of statistical inference. This chapter will explain why the tragic decision taken by the blood donors to die at their own hands might have been premature. For a full understanding of the complexity of this problem, it is sensible to first take a look around the kitchen of science, venture into its laboratories, and smell all the mixtures contained in the test tubes and retorts. Contemporary scientific methods and tools have become so refined that one would have to spend months or even years studying the subject so as to get a proper understanding of the cooking processes that take place in that peculiar kitchen. It is certainly no easy task for the man in the street. I will, however, make the effort to show you, dear reader, around. The game is certainly worth the candle because this is the place where the most enormous frauds can be found, ones that later in a blaze of glory fill pages of scientific journals and books, and appear on shelves in bookstores and libraries. Having promised to take on this challenge, I expect you to give me the same effort and focus in return that will help you sneak into the kitchen of science. Most likely this will be the most difficult of all the chapters in this book; nevertheless, the knowledge that you will gain is worth your trouble. You will understand more than many a scientist who, when working on another experiment, run thoughtlessly through their routine activities to the end. If, however, your aversion to numbers and logical thinking gains the upper hand and outweighs my zeal to explain things to you, have no hesitation in passing over this chapter. This will not prevent you from understanding the ones that follow.

In order to enter and move freely around the laboratory, you should be dressed and look like a scientist. Therefore, don a white lab coat, push your glasses onto the bridge of your nose, and get a bulky book or a bunch of papers to carry under your arm. This is not everything. When approached by another regular practitioner in the kitchen of science, be ready to talk to them in an appropriate manner. Above all, however, you should know where to look so as to see the things of greatest interest. Imagine you are a young, aspiring scientist-psychologist and you are about to start a most exciting project whose outcome may provide an insight into the deepest secrets of human nature...

Scientific research in any field starts from studying the literature available on the chosen subject. After all, nobody wants to investigate things that have already been discovered by someone else. Likewise, the prospect of choosing phenomena that other scientists have already studied and delivered no anticipated outcomes is not particularly tempting. On the other hand, if a subject has already been examined and scientific evidence has been made available, this enables you as a researcher to go a step further in the planned experimentation project. It is critical that you pay close attention at this stage as it is a very important part of the process. If you do, you will acquire more competence to judge the practices of scientists thereafter. Suppose you have noticed, or seem to have noticed, that children from poorer families perceive coins they take in their hand as larger in size than they actually are. You would like to put this hypothesis under scientific scrutiny. Consequently, the problem that lies within your interest is whether the perception of coin sizes by children is determined by whether they are poor or rich. Additionally, suppose that you have not encountered any preliminary research concerning this subject in the literature to date, therefore if you succeed, the odds in favor of making an interesting discovery are high.

In compliance with all standards learnt by future scientists at methodology and statistics classes, it is now time to set up your experiment and carry out your scientific investigation. This should be done so as to verify how likely it is for the pattern you have just observed to represent a possible real difference between groups, and not just an accidental finding. This is a very important criterion in scientific research evaluation. In this chapter you will make yourself acquainted with tools that are designed to effect such an evaluation, ones that are generally known as statistics. But this is still to come, since now what you need to do is prepare and run your own experiment. Off to work then!

The major step in the entire process is to formulate scientific hypotheses for your experiment. In this case, the goal is to examine the impact of affluence on children's perception of coin sizes. Scientists would refer to the children's affluence as an independent variable, that is one that is not contingent on what the children under study will do. They are either poor or rich. Similarly, independent variables include factors that are not impacted by study participants in any way, that is to say such parameters as age or gender, but also the situation set up by the experimenter. The perception of coin sizes represents a dependent variable, because it is assumed to be conditional on those invariable features. Dependent variables can be easily recognized by the fact that they are measured, whereas independent ones are only adjusted, for example, by assigning a study participant to a group of the rich or the poor or by setting up a relevant experimental situation. The formulation of hypotheses itself is relatively easy because it involves determining in what way the independent variable influences the dependent one. In your experiment, the hypothesis may be the following: "Children from a poor background will perceive coins as larger in size than they in fact are." Such a statement is, however, too colloquial to constitute a scientific hypothesis. Noble scientists would sniff at how it is phrased because it embraces a number of assumptions, for instance:

− that the level of affluence of children has an impact on the perception of coin sizes;
− that poor children perceive coin sizes as larger;
− that rich children perceive coin sizes as they actually are;
  however, it is also likely that:
− rich children perceive coins as smaller than they really are;
− or other similar statements.

To avoid such entanglements, scientists are very precise in forming their hypotheses. Interestingly, they assign a kind of opposite statement to each one of their hypothesis, which is commonly known as the null hypothesis – H_o. This is an assumption that asserts no relationship between the independent and dependent variables. The task in any research design is to reject or disprove the null hypothesis. Why is it done in this way? The practice of rejecting the null hypothesis is based on the principle of falsification introduced by one of the most outstanding philosophers of science, Karl Popper. In consonance with his theory, we cannot conclusively affirm a hypothesis, but we can conclusively negate it.[113] The validity of knowledge is tied to the probability of falsification. For illustration, take a very general hypothesis such as "every politician is a liar." This can never be proven since its verification would require testing all politicians in the world for tendencies towards lying. On the other hand, we can falsify this thesis simply by finding politicians who are not liars. For Popper, the scientific method involves "proposing bold hypotheses, and exposing them to the severest criticism, in order to detect where we have erred."[114] If a hypothesis can stand "the trial of fire," then we can confirm its validity. Therefore, hypotheses should be formulated in such a way that they are capable of being proven false.

To get a better understanding of this process, consider the example of court trials and the presumption of innocence principle. When a defendant appears before the court, the null hypothesis is in effect. This means that people are not assumed to be guilty of the crime with which they are charged. In the course of investigation or the court proceedings, the burden is on the prosecutor to show evidence that the defendant is not innocent. Hence, what the prosecutor aims at is negating the null hypothesis and, consequently, putting the defendant behind bars (i.e. validate the alternative hypothesis that claims that the person is guilty). Only if the prosecution has accumulated enough evidence will they succeed in doing so. Otherwise, they are unable to reject the null hypothesis and declare the person guilty.

Now, since it is clear where the practice of formulating null hypotheses stems from, let me help you formulate your hypotheses and simplify the research so as to avoid too much confusion. Your experiment could be grounded on the following null hypothesis:

H_o – the affluence of children does not have any impact on their perception of coin sizes.

By contrast, the proper research hypothesis could state as follows:

H_1 – the affluence of children has an impact on their perception of coin sizes.

The latter will be called the alternative hypothesis. To keep our analysis coherent, let us leave aside other presumptions, such as that poor children perceive coin sizes as larger than they really are.

As you should remember from your methodology classes (just pretend if you do not), a good experimental design should assign participants randomly to two groups. For this purpose, you set up a group of rich children and another of poor children, where the members of the latter come from families with a monthly income of less than 1,000 US dollars, while those of the former have a monthly income over 2,000 US dollars. You show them, or better still, give them coins of different nominal values, for example, one dollar, 50 cents, and other coins, and then ask them to estimate their sizes, for example, by guessing the diameter in inches. Note down all their answers scrupulously and ensure that each child receives the same instruction from you (scientists call this standardization of research). It is also essential to make certain that other children do not hear what the one who is being questioned is saying (scientists call this control over extraneous variables). There you have it! Your experiment has just been completed.

The next step is to calculate the results. How to do this? You may select one of two paths. The first, more difficult one involves the necessity of understanding the statistics you will use, coupled with conduct that complies with descriptions contained in good methodological handbooks. The second, easier path consists in repeating what the majority of psychology researchers do. I will lead you down the second path, because if the majority were to proceed via the first one, there would be no use in writing this chapter. So please do not treat what we will now do as a perfect model.

You have without doubt figured out the mean result for each group because this is what you were taught at your math lessons back at school. What still needs to be examined is whether differences in the results are a matter of chance or whether there is a regular pattern. How can this be tested? At this point, scientists reach for statistical tools known as tests for differences between means. I will not explain how these tools were constructed because doing so would require too much effort both from you and me and, anyway, this knowledge is not crucial for understanding the essence of the problem. As a matter of fact, just between us, many scientists have little if any grasp of these methods, too. Whereas several years ago such statistics were kept manually and with the use of a calculator at best, nowadays the most laborious task is to enter the data into the system, and professional software designed to perform statistical data analysis does the rest. Scholars put a lot of trust in such applications, which do the hard work for them. Once your data has been entered into the system, you will get the results in a flash. Suppose that you instruct the computer to calculate the differences between the means through the student's t-test, which is a statistical hypothesis testing device well-liked by scientists. The value that should be of your greatest interest will be marked with a small letter p. If this number is 0.05 or less, you have every reason to be optimistic because it means that in your research has achieved the statistically significant result that is accepted by most scientific journals. It is hence safe to say that the difference between the means in two groups is not due to any random factors, but is a statistically significant result. To be more accurate, your hypothesis has just been confirmed and the probability of error amounts to 5% (or less). Thus, you are now therefore free to set about writing an article, then send it for review and have it published in one of the many renowned journals, thereby proving that the perceived size of coins is impacted by the level of children's affluence.

For all that, reader, you are perhaps curious why I have not pointed to a few other numbers right in front of your eyes, be it on the computer screen or on the statistical software printout. If so, take it from me that your curiosity is indeed exceptional. Most of the time, scientists pay no attention to such details. When asked about the meaning of these other figures, they will claim that in the interpretation of research nothing counts more than simply the level of significance. It is critical to obtain the minimum of p-value less than the significance level (denoted alpha α). A significance level is chosen before data collection and is often set to 0.05 (5%). It is accepted by most journals, because it implies 5% or less probability of error. The lower the p-value, the better. What else should one be concerned about? Until quite recently, this viewpoint has been common for the bulk of editors and reviewers from the most prestigious journals, and those of a lower standard would just follow in their footsteps. Is the majority right? I am afraid not and, sadly, the overwhelming number of psychological studies that have been published for decades seem to be simply ... useless.

In 1962, Jacob Cohen, who is known to thousands of research psychologists as the inventor of the kappa statistic (often called Cohen's kappa) and the popularizer of multiple regression analyses in psychological research, published an article entitled "The Statistical Power of Abnormal-Social Psychological Research"[115] in the *Journal of Abnormal and Social Psychology*. He put forward an analysis of 70 research studies that reported a significance level of 0.05 issued in the years 1960–61 in *Journal of Abnormal and Social Psychology*. The findings were shocking, to say the least. The average level of error, it turned out, amounted to more than 50%, and not 5%, as had been widely assumed! A result that is even higher than a flip of a coin would give. How did it come to be that the error levels in the analyzed studies were so monstrous? To understand this, we need to take a step further in our analysis of research methods.

Let us return to the hypotheses that we previously established together, the null and alternative hypotheses. When conducting research and drawing inferences about relationships between the variables, we take decisions with regard to the hypotheses. In doing so, we expose ourselves to two types of errors. For instance, we may reject a null hypothesis that in fact is true. Scientists define this decision as a Type I error (false positive). Estimation of the probability of committing this type of error is marked with the symbol α and referred to as the test significance level. A Type II error (false negative) is a decision to retain the null hypothesis that is in fact false. The probability of committing the error is marked with the symbol β (the Greek small letter for beta) and the difference between the number one and this probability value (1-β) is called *statistical power*.

Those who teach methodology speak also about a Type III error that is also quite common and occurs when the two errors are ... conflated with each other. When examining the impact of affluence on the perception of coin sizes, what we are in fact testing is solely the following null hypothesis: that affluence does not have an impact on the perception of coin sizes. This hypothesis is subsequently nullified as untrue or we state there are no grounds for rejecting it. Perhaps you may be surprised at the fact that it is the null hypothesis that becomes the center of attention. Indeed, this is just what scientists do to remain in line with the previously mentioned falsification principle. Rejecting the null hypothesis is assumed to be tantamount to acceptance of the fact that predictions stipulated by the alternative hypothesis occur more frequently than by sheer chance. The table below shows that there are four alternatives of decisions that can be made with regard to the null hypothesis:

	H_o true	H_o false
Reject H_o	Type I error	Correct decision
Retain H_o	Correct decision	Type II error

A story written by Chong Ho Yu, a psychometrician and an expert in digital media instructional technology, should help us connect the dots between both errors and their consequences:

> Once a warship is patrolling along the coast. Suddenly an unidentified aircraft appears on the radar screen but the computer system is unable to tell whether it is a friend or a foe.
> The captain says: The null hypothesis is that the incoming aircraft is not hostile. If it is indeed hostile and I don't fire the missile, it is a Type II error. The consequence of committing this Type II error is that we may be attacked and even killed by the jet.
> The alternate hypothesis is that the incoming aircraft is hostile. But if it is not hostile and I shoot it down, it is a Type I error. The consequence of making this Type I error is the termination of my career in the Navy.
> It seems that the consequence of Type II error is more serious. Therefore, I disbelieve in the null hypothesis. Fire!

The commander shouts "Delay the order!" He argues: If the null hypothesis is false but we don't react, the consequence is that a few of us, let's say 30, may be killed.

If the alternate hypothesis is false and actually the incoming aircraft is a commercial airliner carrying hundreds of civilian passengers, the consequence of committing a Type I error is killing hundreds of innocent people and even starting a war that may eventually cause more deaths.

I assert that the consequence of Type I error is more severe. Thus, I disbelieve in the alternate hypothesis. Hold the fire!

The above story is exaggerated to make this point: Subjective values affect balancing of Type I and Type II error and our beliefs on null and alternate hypotheses. A similar scenario could be seen in the movie "Crimson Tide" and two real life examples that happened in 1987 and 1988. In 1987 an Iraqi jet aircraft fired a missile at the USS Stark and killed 37 US Navy personnel. A patrol plane detected the incoming Iraqi jet and sent the information to the USS Stark, but the Captain did not issue a red alert. A year later the USS Vincennes patrolling at the Strait of Hormuz encountered an identified aircraft. This time the Captain ordered to open fire but later it was found that the US warship shot down an Iranian civilian airliner and killed 290 people. While the former mistake is caused by under-reaction, the latter is due to over-reaction.[116]

Following this example, the consequences of committing both types of error, say, in the process of testing new drugs can be easily envisaged. The null hypothesis could be that the examined drug causes no dangerous side effects. If it indeed does, and it is not withdrawn from production, it is a Type II error. The consequence of committing a Type II error is that many people could suffer from a range of adverse health outcomes. The alternate hypothesis is that the examined drug causes side effects, which of course can be dangerous for people's health. But if it is not dangerous and still its production is discontinued, it is a Type I error. The consequence of making a Type I error in this case is that many very ill people are deprived of a needed drug, and this in turn may lead to their deaths.[117]

Likewise, a similar situation arises when a diagnostic test of a serious disease is made. The null hypothesis could be that the examined person is not ill. If they are actually ill and I, as their doctor, do not implement a relevant treatment plan, it is a Type II error. The consequence of committing this Type II error is that the patient in question could eventually die. The alternate hypothesis is that the examined person is ill. But if they are not and the treatment is effected nevertheless, it is a Type I error. The consequences of making this Type I error are possible health hazards connected with this therapy.

Severe consequences could also take place in court when a mental health professional is asked to determine the defendant's sanity, that is, to judge whether they can be held liable for their behavior (whether the person is of sound mind). The null hypothesis could be that the examined person is liable for their actions (is of sound mind). If they indeed are not liable (of unsound mind) and no isolation order is issued, it is a Type II error. The consequence of committing this Type II error is that this person could pose a danger to other people. The alternate hypothesis is that the examined person is not liable (is of unsound mind). Yet if they are not, and an isolation order is issued, the consequence is that an innocent person goes to jail. This is a Type I error.

A number of other situations can be conceived of to illustrate the more or less severe consequences of both types of error. Whenever a researcher rejects a null hypothesis that is in fact true, they commit a Type I error and the probability of committing this type of error is equal to α. If, however, the researcher accepts the null hypothesis that is actually false, a Type II error is made, and the probability of this type of error equals β. Still, reader, you are doomed to failure if you attempt to search for this beta in most of the empirical studies in psychology. It is absent from the better part of them, though, it should be admitted, it has appeared in some research reports more frequently in recent times and major journals are beginning to request its value be specified by researchers. This approach to hypothesis testing, in which only the probability of Type I error is allowed, has by tradition established itself in the field of psychology. John Hunter from Michigan State University, known among psychologists for his work on methodology, was very direct about this situation:

> The significance test as currently used is a disaster. Whereas most researchers falsely believe that the significance test has an error rate of 5%, empirical studies show the average error rate across psychology is 60% – 12 times higher than researchers think it to be. The error rate for inference using the significance test is greater than the error rate using a coin toss to replace the empirical study. The significance test has devastated the research review process. Comprehensive reviews cite conflicting results on almost every issue. Yet quantitatively accurate review of the same results shows that the apparent conflicts stem almost entirely from the high error rate for the significance test. If 60% of studies falsely interpret their primary results, then reviewers who base their reviews on the interpreted study "findings" will have a 100% error rate in concluding that there is conflict between study results.[118]

Where the 60% error rate comes from will be discussed elsewhere in this chapter. Before we move onto this, it is important that you know that many researchers are not even aware of the enormity of this error – those shall be called ignoramuses. Some of them realize how low the quality of papers that get published is, but they are not particularly troubled by this fact. This attitude reveals a particular air of nonchalance that is simply inexcusable. There is also a group of those who just feel more comfortable with this approach and have no real interest in any changes or impediments. Many editors and reviewers who, being aware where the flaws are hidden, do not verify their requirements, certainly belong to this category. They should be accused of acting intentionally to the detriment of science. In his article, John Hunter provides a good illustration of these frames of mind by quoting responses of psychologists to the news of the 60% error rate. The ignoramuses would give it no credence, saying: "A 60% error rate is impossible; the error rate is only 5%."[119] Those affected by nonchalance reflect: "This is no surprise. Everyone knows that far more than 60% of studies done in psychology are garbage studies with major methodological errors. However, my studies are methodologically top notch and so my studies do not make errors."[120] Unfortunately, these comments are also replete with statements that prove that even among scientists there will be those who deliberately miss the truth: "You're right, but there is nothing we can do about it. I know that the significance test has a 60% error rate, but I'm not going to tell an editor that. I include significance tests in all my papers because they'll get rejected if I don't."[121]

An alternative approach to research methodology that concentrates on Type II errors (β) as well as on statistical power ($1-\beta$) arose within the framework of a statistical theory developed by Jerzy Neyman, a mathematician of Polish origin, and Egon Pearson, a British statistician. It was, however, sharply criticized by well-known geneticist and statistician Ronalda Fisher who, at the same time, was the author of the null hypothesis significance testing (NHST – because this is how the method is referred to by scientists) discussed in this chapter. In critical comments with regard to statistical power, he went as far as to compare Neyman and Pearson to "Russians, who were trained in technological efficiency as in a 5-year plan, rather than in scientific inference."[122] In response to such attacks and in the light of the statistical power of conducted research, Neyman described some of the methods proposed by Fisher as "worse than useless" in a mathematically specifiable sense.[123] The discussion between the statisticians that progressed in the 1950s did not draw enough attention from psychologists. Fisher's methods had worked well in agronomy as it had originally been the field of their application. In psychology, they disappointingly gave birth to research that at the level of statistical significance of $\alpha \leq 0.05$ reports relationships of much lower quality than would be indicated by the coin toss, as Hunter suggested. Let us look into specific numbers to get to the bottom of this matter. We will go over the process of making a diagnosis, which in its essence is nothing else than the procedure of null hypothesis testing itself.[124]

The real numbers in practice are different from these analyzed below, but for the clarity of thinking let's assume the following. Schizophrenia among adults is observed in about 2%* (I have added * symbols next to the letters for easy reference in the tables that follow) of the population. By implication, the actual probability that a screening test will identify a non-schizophrenic amounts to as much as 98%**. The test recommended to detect schizophrenia is of minimum 95%*** sensitivity, understood as its ability to identify those who are in fact affected by the disorder out of all persons with the condition under study (i.e. patients who suffer from schizophrenia). If, therefore, we test a group of patients, its sensitivity will inform us what percentage of this group had a positive result. In our case the possible rate of error equals 5%****. At the same time, the accuracy of the test in identifying patients as disease-free is around 97%*****. This test property is called specificity and it refers to the ability of the test to correctly detect individuals without the disease. Hence, if we examine a group of non-schizophrenic individuals, test specificity informs us what percentage of them has a negative test result. For this reason, the likelihood of a healthy person being correctly identified as not having the disease amounts to 97%, and the possible error rate is 0.03%******. The likelihood that a person suffering from schizophrenia will be actually considered as schizophrenic is in turn 95%***, and the diagnosis error rate is 5%****.For clarity, be mindful of the fact that if diagnostic purposes are considered, we adopt the following two hypotheses:

H_0 – the examined person is non-schizophrenic

H_1 – the examined person is schizophrenic

With a test result that has identified you as schizophrenic with 95% accuracy, the probability that actually you are not is less than 5%. Based on this rate of statistical significance, a researcher will most likely reject the null hypothesis as untrue and you yourself, being diagnosed with schizophrenia, will undoubtedly end up in a mental hospital. Do not let this upset you too much though, and reflect how impulsive the blood donors from Florida were to end their lives so tragically, without first calculating the real probability of their HIV infection. Let us see where the actual error might have been committed by the person doing the test.

Though it may seem so at first glance, given a positive test result for schizophrenia (Type I error probability), the chance that a case is actually non-schizophrenic is not the same as has been determined above. It is neither true that the probability that a case identified as schizophrenic is really healthy amounts to 5%, nor that the probability of such an inaccurate diagnosis is quite low. In fact, the chance the case does not have schizophrenia even if the test has indicated it is around 60%! Let us look into it in a more methodic manner.

Suppose that we subjected 1,000 individuals to the test following the approach adopted by researchers who test the null hypothesis, that is, focusing primarily on whether we should accept or reject it. By way of reminder, our null hypothesis is that the case is non-schizophrenic. The hypothesis will be in fact false with regard to 20 individuals who really suffer from schizophrenia (2% schizophrenics within the population) and true for 980 non-schizophrenic individuals (see the bottom row of the table).

Our test enables us to detect with 97% accuracy non-schizophrenics within a sample of 980 non-schizophrenic individuals, so this number should run to 950. The remaining approximate 30 results from 3% error (see the first column). At the same time, we have a 95% chance of identifying schizophrenics within the population of 20 who actually have schizophrenia, which accounts for 19 individuals. One individual with actual schizophrenia will receive a false negative test result (see second column).

Mathematical operations on such large numbers naturally involve fractions and since it is impossible to consider people in fractional values, for example, as 29.4 individuals, the data has been rounded up to one person.

The null hypothesis: the case is non-schizophrenic.

	H_0 true (Actual non-schizophrenic individuals)	H_0 false (Actual schizophrenic individuals)	**Total:**
Rejection H_0 (positive test result)	30 cases incorrectly identified as schizophrenic **Type I error** (3%***** out of 980)	19 cases (95%*** out of 20)	49
Retention H_0 (negative test result)	950 cases correctly identified as non-schizophrenics (97%***** out of 980)	1 case with actual schizophrenia identified as non-schizophrenic **Type II error** (5%**** out of 20)	951
Total:	980 (98%** out of 1000)	20 (2%* out of 1000)	

And now let us compute what the actual Type I error is. We have rejected the null hypothesis with regard to 49 cases. How many false positive cases of schizophrenia are there in this group? Already at first glance it is apparent that more than half! The exact number is 61.2%. At this point it should be more apparent that the blood donors from Florida need not have acted so hastily when departing from this world voluntarily. In the same way, you have presumably understood why I told you not to worry about your test result. It would be worthwhile if this example made you aware of the most profound parallel. Researchers who conduct experiments and make inferences based on $\alpha \leq 0.05$ act in the manner identical to that of diagnosticians in the schizophrenia example.

There is one more thing you need to know. The statement that the level of statistical significance is equivalent to the level of committed errors is only true if the null hypothesis itself is true. In our example this would limit our considerations to the left-hand side of the table. Nevertheless, most research studies reject the null hypothesis as true, and that being the case, apart from the probability of committing a Type I error, we must also factor in the probability of a Type II error, i.e. calculate the value of β. This is what most researchers fail to do, and that is why this distorted and ill-designed process of proving something has been recognized in the social sciences as the most valid and effective for so long, although lone-wolf critics have for years exhorted the profession to use alternative methods designed for hypothesis testing. It was just upon the inclusion of the probability of the Type II error that Cohen, Hunter, and other researchers named in this chapter arrived at the 60% or higher probability of obtaining erroneous results. Hunter has gone even further and mentioned an error close to the maximum that can be achieved for the conventional two-tailed test (i.e. analysis of variance), which is 97.5%.

These errors could be avoided without any trouble by simply working out the values of confidence intervals and the statistical power of the study, as well as by analyzing effect size or reaching for Bayesian statistics (due to the nature of this book, I will not walk you through these methods). However, cargo cult priests are not easily talked into changing their rituals. More than half a century ago, Jacob Cohen started a determined campaign against NHST. In one of his articles he argued that "we, as teachers, consultants, authors, and otherwise perpetrators of quantitative methods, are responsible for the ritualization of null hypothesis significance testing (NHST; I resisted the temptation to call it statistical hypothesis inference testing) to the point of meaninglessness and beyond. I argue herein that NHST has not only failed to support the advance of psychology as a science but also has seriously impeded it."[125] Cohen persisted in his efforts until his death in 1998. Joseph Rossi, professor of psychology at the University of Rhode Island, investigated whether Cohen's struggle brought about any positive results, based on an analysis of three major journals from 1982. His findings were plain and simple: "20 years after Cohen (1962) conducted the first power survey, the power of psychological research is still low."[126]

Several years later, a pair of German psychologists from the University of Konstanz, Peter Sedlmeier and Gerd Gigerenzer, carried out an analysis that was expected to answer the title question in their paper "Do Studies of Statistical Power Have an Effect on the Power of Studies?" In its summary they concluded, "The long-term impact of studies of statistical power is investigated using J. Cohen's (1962) pioneering work as an example. We argue that the impact is nil; the power of studies in the same journal that Cohen reviewed (now the *Journal of Abnormal Psychology*) has not increased over the past 24 years."[127]

Jacob Cohen himself spoke on this in the mid-1990s through his article "The Earth Is Round *(p < .05)*." His straightforward conclusion left no doubt that we are invariably locked in the cage of a cargo cult: "After 4 decades of severe criticism, the ritual of null hypothesis significance testing – mechanical dichotomous decisions around a sacred .05 criterion – still persists."[128] Before his death, Cohen was able to address the subject for the last time when responding to critical reviews of the article above.[129]

Although there has been a steady growth in the use of Bayesian statistics, which constitute an alternative tool for computing actual values of effects in research, many scientists, including from outside psychology, continue to promote the ritual of NHST. In 1999, a political scientist from Cal Poly University, Jeff Gill, published an analysis that shows how political scientists draw on useless NHST on a widespread basis.[130]

Things seemed to fare slightly better in health psychology. Psychologists Jason Maddock and Joseph Rossi calculated statistical power $(1-\beta)$ for 8,266 statistical tests in 187 journal articles published in the 1997 volumes of *Health Psychology*, *Addictive Behaviors*, and the *Journal of Studies on Alcohol*. They concluded as follows: "J. Cohen (1988) recommended that the power to detect effects should be approximately .80. Using this criterion, the articles in these journals have adequate power to detect medium and large effects. Intervention studies have much less power to detect effects than nonintervention studies do. Results are encouraging for this field, although studies examining small effects are still very much underpowered. This issue is important, because most intervention effects in health psychology are small."[131]

Regrettably, despite recommendations issued by the APA and unhurried changes in the practices of some researchers, textbooks used in education research methods and statistics classes almost always fail to acknowledge that there is controversy surrounding NHST. This is what has emerged from the analysis carried out by Jeffrey Gliner and colleagues, who examined 12 recently-published textbooks.[132]

Patrizio E. Tressoldi and colleagues decided to verify whether journals that have high impact factors applied higher statistical standards. They made an evaluation of all articles related to psychological, neuropsychological, and medical issues published in 2011 in four high-impact journals: *Science*, *Nature*, *The New England Journal of Medicine*, and *The Lancet*, as well as three journals with comparatively lower impact factors: *Neuropsychology*, the *Journal of Experimental Psychology-Applied*, and the *American Journal of Public Health*. Their findings offer no cause for optimism. NHST, deficient in confidence intervals, effect size, prospective power and model estimation, proves to be the prevalent statistical practice identified in articles published in *Nature* (89%), followed by articles published in *Science* (42%). By contrast, in all other journals, both with high and lower impact factors, most articles report confidence intervals and/or effect size measures. The authors interpreted these differences as a manifestation of consequences of editorial policies adopted by the journal editors.[133]

Note that I am writing this more than half a century after distortions inherent to NHST were first exposed. Although much has changed in this field today, an array of editorial practices remains far from perfection. As Marc Branch, psychologist from University of Florida, emphasizes, "Six decades-worth of published information has shown irrefutably that null-hypothesis significance tests (NHSTs) provide no information about the reliability of research outcomes. Nevertheless, they are still the core of editorial decision-making in psychology."[134]

Meanwhile, in 2014, Geoff Cumming, a preeminent author in the field of statistics, while looking through contributing factors behind the incompleteness and untrustworthiness of psychological research literature, concluded that: "in response to renewed recognition of the severe flaws of null-hypothesis significance testing (NHST), we need to shift from reliance on NHST to estimation and other preferred techniques."[135]

Now, reader, having discovered such misuses of scientific methodology, take a walk to a library or a bookstore to browse through psychological textbooks. When you are doing this, reflect that most of the scientific content you see is based on NHST. What is more, the lion's share of knowledge psychology students are taught is based on NHST. Even most of what scientists discuss at all those portentous conferences and congresses or in their professional journals is based on NHST. Why is this so? Well, the answer is right under your nose, indicated by Feynamn's remark: because ritual, not reality, is what cargo cult science and cargo cult priests are concerned about. At this point would be advised to recall the words of the economist and historian Deirdre McCloskey: "scientists care about whether a result is statistically significant, but they should care much more about whether it is meaningful."[136]

---

[108] G. Gigerenzer, *Calculated Risks: How to Know when Numbers Deceive You* (New York: Simon and Schuster, 2002), 120.
[109] Ibid., 121.
[110] Ibid., 121.
[111] G. J. Stine, *Acquired Immune Deficiency Syndrome: Biological, Medical, Social, and Legal Issues* (Englewood Cliffs, NJ: Prentice Hall, 1996), 333, 338. As cited in Gigerenzer, *Calculated Risk*, 121-122.
[112] The cited data and facts concern the 1908s, and should not be invoked when assessing the manner in which HIV testing and diagnosis are conducted at present.
[113] K. R. Popper, *Logic of Scientific Discovery* (London: Hutchinson, 1959).
[114] K. R. Popper, "Replies to my Critics," in *The Philosophy of Karl Popper*, ed. P. A. Schilpp (La Salle: Open Court, 1974), 963-1197, 68.
[115] J. Cohen, "The Statistical Power of Abnormal-Social Psychological Research." *Journal of Abnormal and Social Psychology 65*, (1962): 145-153.

[116] C. H. Yu, "Don't Believe in the Null Hypothesis?" http://www.creativewisdom.com/computer/sas/hypothesis.html

[117] In fact, clinical research on the efficacy of drugs and their side effects are far more complicated. I used this example only for the purpose of illustrating the consequences that result from committing both types of errors.

[118] J. E. Hunter, "Needed: A Ban on the Significance Test." *Psychological Science 8*, (1997): 3-7, 3.

[119] Ibid., 3.

[120] Ibid., 3.

[121] Ibid., 4.

[122] R. A. Fisher, "Statistical Methods and Scientific Induction," *Journal of the Royal Statistical Society 17*, (1955): 69-78, 70.

[123] W. Stegmüller, *"Jenseits von Popper und Carnap":Dielogischen Grundlagen des Statistischen Schliessens* (Berlin: Springer-Verlag, 1973).

[124] J. Cohen, "The Earth is Round (p<.05)," *American Psychologist 49*, (1994): 997-1003.

[125] Ibid.

[126] J. S. Rossi, "Statistical Power of Psychological Research: What we Have Gained in 20 Years?" *Journal of Consulting and Clinical Psychology 58*, (1990): 646-656, 646.

[127] P. Sedlmeier, and G. Gigerenzer, "Do Studies of Statistical Power Have an Effect on the Power of Studies?" *Psychological Bulletin 105*, (1989): 309-316, 309.

[128] Cohen, "The Earth," 997.

[129] J. Cohen, "The Earth is Round (p<.05): Rejoinder," *American Psychologist 50*, (1995): 1103.

[130] J. Gill, "The Insignificance of Null Hypothesis Significance Testing," *Political Research Quarterly 52*, (1999): 647-674.

[131] J. E. Maddock, and J. S. Rossi, "Statistical Power of Articles Published in Three Health Psychology-Related Journals," *Health Psychology 20*, (2001): 76–78.

[132] J. A. Gliner, N. L. Leech, and G. A. Morgan, "Problems With Null Hypothesis Significance Testing (NHST): What Do the Textbooks Say?" *The Journal of Experimental Education 71*, (2002): 83–92.

[133] P. E. Tressoldi , D. Giofré, F. Sella, and G. Cumming, "High Impact=High Statistical Standards? Not Necessarily So" *PlosONE*, (2013), http://journals.plos.org/plosone/article? journal.pone.0056180id=10.1371/

[134] M. Branch, "Malignant Side Effects of Null-Hypothesis Significance Testing." *Theory & Psychology 24*, (2014): 256-277.

[135] G. Cumming, "The new Statistics: Why and how." *Psychological Science 25*, (2014): 7-29.

[136] D. N. McCloskey, "The Insignificance of Statistical Significance." *Scientific American 272*, (1995): 104-105.

Chapter 5: Neuroscience: From Phrenology to Brain Porn

The Fowlers were a family of farmers from New York with three children - Lorenzo Niles, Orson Squire, and Charlotte. They all worked hard from a very young age. At the beginning of the 19th century, boys would typically be attracted to a career in the priesthood, so it was therefore no surprise when Lorenzo, having received a diploma from a district school, undertook studies at the Amherst Academy to become a clergyman. At the same time, his elder brother, Orson Squire, who was deeply interested in the new science of the mind, became a convert to phrenology mainly due to the influence of fellow students and the writings of fashionable promoters of the subject, such as Johann Spurzheim and George Combe. Lorenzo followed in his footsteps and soon after, assisted by their sister Charlotte, they started their own business in phrenology, which consisted in reading heads and offering lectures on the subject. The new science of the mind, not unexpectedly, turned out to be a fairly lucrative field and hence the Fowlers' initial idea of becoming men of the cloth fell by the wayside. The brothers were so dedicated to their work that soon, in 1836, Lorenzo set up a phrenological establishment in New York, and two years later another one was launched in Philadelphia. The Fowlers also founded the *American Phrenological Journal* and *Miscellany American Phrenological Journal*. They expanded into publishing, producing reprints of such phrenological greats as Spurzheim and Combe, as well as their own writings. By the 1840s, they had become owners of one of the largest publishing concerns in New York, and their phrenological services were highly valued and sought by the public.

Lorenzo travelled to England where he lectured widely and his tours led to the creation of new phrenological societies, just as earlier lecturers like Spurzheim and Combe had inspired similar societies decades before. Fowler also made an appearance at the International Exhibition of 1862. He personally, along with members of his family and his employees, successfully gave practical instruction on phrenology, provided readings and character analyses, and offered weekly courses on practical phrenology. They also sold voluminous literature (their publications becoming somewhat ubiquitous in the field) and phrenological busts. The Fowlers made a name for themselves as professional practical phrenologists who read thousands of heads for a fee, actively lectured, and also wrote extensively on the subject.

One day in the early 1870s, a man came to Lorenzo's office to have his head read. This is how he recounted Lorenzo's examination:

> I found Fowler on duty, in the midst of the impressive symbols of his trade. On brackets, on tables, on shelves, all about the room, stood marble-white busts,

hairless, every inch of the skull occupied by a shallow bump, and every bump labeled with its imposing name, in black letters.
Fowler received me with indifference, fingered my head in an uninterested way and named and estimated my qualities in a bored and monotonous voice. He said I possessed amazing courage, an abnormal spirit of daring, a pluck, a stern will, a fearlessness that were without limit, I was astonished at this, and gratified, too; I had not suspected it before; but then he foraged over on the other side of my skull and found a hump there which he called "caution." This bump was so tall, so mountainous, that it reduced my courage-bump to a mere hillock by comparison, although the courage-bump had been so prominent up to that time – according to his description of it – that it ought to have been a capable thing to hang my hat on; but it amounted to nothing, now, in the presence of that Matterhorn which he called my Caution. He explained that if the Matterhorn had been left out of my scheme of character I would have been one of the bravest men that ever lived – possibly the bravest – but that my cautiousness was so prodigiously superior to it that it abolished my courage and made me almost spectacularly timid. He continued his discoveries, with the result that I came out safe and sound, at the end, with a hundred great and shining qualities; but which lost their value and amounted to nothing because each of the hundred was coupled up with an opposing defect which took the effectiveness all out of it.
However, he found a cavity, in one place; where a bump would have been in anybody else's skull. That cavity, he said, was all alone, all by itself, occupying a solitude, and had no opposing bump, however slight in elevation, to modify and ameliorate its perfect completeness and isolation. He startled me by saying that that cavity represented the total absence of the sense of humor! He now became almost interested. Some of his indifference disappeared. He almost grew eloquent over this America which he had discovered. He said he often found bumps of humor which were so small that they were hardly noticeable, but that in his long experience this was the first time he had ever come across a cavity where that bump ought to be.[137]

The person with the diagnosed total absence of the sense of humor was none but Mark Twain himself, who had visited Fowler under an assumed name. It was not, however, his last encounter with the famous phrenologist. As he reported:

I was hurt, humiliated, resentful, but I kept these feelings to myself; at bottom I believed his diagnosis was wrong, but I was not certain. In order to make sure, I thought I would wait until he should have forgotten my face and the peculiarities of my skull, and then come back and try again and see if he had really known what he had been talking about, or had only been guessing. After three months I went to him again, but under my own name this time. Once more he made a striking discovery – the cavity was gone, and in its place was a Mount Everest – figuratively speaking – 31,000 feet high, the loftiest bump of humor he had ever encountered in his life-long experience![138]

How have things changed since the famous writer last visited the famous phrenologist? Not much, but for the quite obvious fact that Mark Twain is no longer with us. There is perhaps one difference: nowadays, it is much more common for big names rather to support pseudoscience than think critically and engage themselves in debunking it. Well-known athletes are keen to sport bogus wristbands, the manufacturers of which make firm claims as to their potential maximizing effect. There are many celebrated figures taking active part in the anti-vaccine movement or among opponents of genetically modified food (GMO). Pseudoscience finds its supporters even among royals.

Just like one-and-a-half centuries ago, today there are also many who offer scientifically sound contemporary head reading services. One of them is Daniel Amen, a psychiatrist trained at the Walter Reed Medical Center. He has found fame as the author of several best-selling books and a motivational speaker. He has built his reputation and fortune on performing brain scans and then wildly exaggerating their diagnostic and prognostic worth. His clinics make use of an imaging modality called SPECT, a largely obsolete technology with terrible spatial resolution that has been superseded by PET and MRI for most clinical and research purposes. The only real advantage of SPECT is that it is far less expensive to perform than either PET or MRI, which seems to be the reason why it is still in use. Here is Daniel Carlat, a psychiatrist from Tufts University School of Medicine, on the examination performed by the Fowler of today:

> "So here's your brain," the doctor says, as the center of my mental life pirouettes before me, rendered in electric blues and reds. Daniel Amen, MD, manipulates the screen image with a few taps on his keyboard.
> "It looks good, pretty symmetrical. Red means more activity, blue means less."
> We're peering at a Spect scan taken a half hour ago. He takes a closer look. ...
> "The only question I'd ask you is whether you've ever had a brain injury, because there is low activity in your occipital cortex and your parietal lobe, all on the left side."
> I admit to the occasional fall while snowboarding, but I've always worn a helmet. Amen shakes his head. "Your brain is 80 percent water and the consistency of tofu, and your skull is hard, so your brain was not meant to snowboard, even with a helmet. I recommend tennis or Ping-Pong."
> He calls up a different view, this one from below, as though looking up from the spinal cord. I see a spot on one side that is conspicuously ... empty. "What's that?" I ask.
> "That's a left temporal lobe ding. It's in a fairly innocuous area, but I'd still ask your wife how your temper is."
> ...
> Following my second scan, taken immediately after I concentrate on performing several assigned tasks at a computer, I am ushered into Amen's spacious office.
> ...
> "Today we'll look at both your resting scan and your concentration scan," he says, his expression friendly. "But first, I'd like to go over some of your history."

Amen runs through a series of questions I'm accustomed to asking my own patients. I find him to be an excellent psychiatrist – focused, compassionate, patient – and I feel comfortable revealing some of the more difficult truths of my past. He asks me whether I have a history of psychiatric problems. Yes, I suffered a short bout of depression a few years ago and treated myself – successfully – with the antidepressant Celexa. Do I have any medical problems? Nothing significant. Do I have a history of mental illness in my family? Unfortunately, my mother suffered severe depression and committed suicide when I was in college. After another 15 minutes of questions and conversation, he says, "Let's look at your scans." He takes the images that he printed out this morning and puts them side by side on a large table. He points to several views of the surface of my brain. "What I see here is that activity in your prefrontal cortex is low at rest but becomes better when you concentrate, and your thalamus becomes more active, too. I think this means you have a predisposition to depression."

I nod. Scrutinizing the scans some more, he says, "You need to be busy to be happy. Your brain is cool at rest. You need stuff in your life to feel alive, together, and connected." He looks at another view, this one showing only the most active regions of my brain. "In this scan, you have increased activity in your thalamus, your two basal ganglia, and your cingulate cortex." He picks up a pen and draws a line connecting these four regions to the right lateral temporal lobe. "I call this 'the diamond plus.' It's a pattern of angst, and we see it in people who have had significant trauma in their lives."

He puts down his pen and turns to me. "I would love to see your brain healthier, because you'll be happier if it's healthier," he says. "It's too low in activity. I recommend a multivitamin, and to get better blood flow I would take gingko." Just before I leave, he advises me to lay off the snowboarding and play more tennis. "With the lowered activity in your cerebellum," he explains, "I'd like to see you do more coordination sports." [139]

This evaluation, done in Newport Beach, California, in one of four Amen Clinics, cost 3,300 US dollars. The price included two SPECT scans and a series of clinical interviews. Carlat also received a report on his mental health, along with recommendations as to lifestyle changes, supplements, and medications – a prescription for a "better brain."

Neuroimaging is not broadly used in psychiatry. Most insurance companies will cover a PET scan only if it is used to distinguish Alzheimer's disease from a rare form of dementia. There are only a handful of psychiatrists who would claim the technique is suitable for use in daily clinical care. And still, against all logic, contemporary Fowlers hang on, successfully developing their businesses in the same way as the famous brothers did one-and-a-half centuries ago. The only thing that stands apart is the technological instrumentarium.

Neuroscience is the scientific study of the nervous system and was formerly seen as a branch of biology. However, with the passage of time it has grown into an interdisciplinary science that collaborates with a range of other fields, such as cognitive science, computer science, medicine (including neurology), philosophy, psychology, and many more. Many experimental and cognitive psychologists are now referred to as cognitive neuroscientists.

At lightning speed contemporary neuroscience has become a new buzzword, both among academics and those outside scientific institutions. The other, darker side of the coin is that it has equally quickly given rise to unrealistic expectations and overblown promises of breakthroughs and personal improvement across a virtually unlimited number of areas. Crude simplifications and half-truths that attempt to define this complex field of brain research are commonly found everywhere, from specialist publications to the daily press. The market is saturated with an array of products and services claimed to be grounded in or validated by neuroscience. People can buy neurodrinks or neurosupplements to enhance their physical strength or boost stamina. It is no longer in fashion to market products, but rather to neuromarket them. We can delve into achievements of the neuroeconomy or neurolaw. It is possible to buy neuroart and to order a neurodesign from a designer. Although it has not yet been scientifically substantiated that brain scans can be successfully used for diagnosis, many labs offer neurodiagnostic services. Having received (or purchased) such a diagnosis, one may find neurocounseling or neurotherapy in the blink of an eye. Those who take pleasure in fitness and exercise may now try out neurosport services. Countless web pages created by neuroenthusiasts of neurobabble offer mantra-like promises such as "With the right tools, the possibilities are limitless."

Neuroscience is becoming increasingly widespread in marketing. A whole host of good examples was given by Molly Crockett in her lecture at the TED Conference in London in 2012:

> Here's a study published by a team of researchers as an op-ed in *The New York Times*. The headline? "You Love Your iPhone. Literally." It quickly became the most emailed article on the site. So how'd they [the researchers] figured this out? They put 16 people inside a brain scanner and showed them videos of ringing iPhones. The brain scans showed activation in a part of the brain called the insula, a region which is linked to feelings of love and compassion. So the researchers concluded that because they saw activation in the insula, this meant the subjects loved their iPhones. The main problem with this line of reasoning is that the insula does a lot. It is involved in positive emotions like love and compassion, but it's also involved in many other processes, like memory, language, attention, even anger, disgust and pain. Based on the same logic, one could equally conclude you hate your iPhone. When you see activation in the insula, you can't just pick and choose your favorite explanation from off the very long list. My colleagues Tal Yarkoni and Russ Poldrack have shown that the insula pops up in almost a third of all brain imaging studies that have ever been published. So chances are really, really good that your insula is going off right now, but I won't kid myself to think this means you love me.[140]

Some authors dare to call the ubiquitous neuroscience brain porn.[141] And they are not blasphemers. Researchers proved that the image of the brain can change our judgements in a similar way as sex images do. In one experiment, David McCabe from Colorado State University and Alan Castel from the University of California asked 156 students to read three different brief press articles about brain imaging studies. All of them were fake. Some articles had no image, some had a bar graph, and some had an image of a brain. The articles were about three different topics, but an equal number of students saw each article with text only, the graph, or the brain image. Then the students were asked several questions. In one of them, they had to evaluate whether the scientific reasoning in the article made sense. The articles accompanied by brain images were rated significantly higher than the other articles, despite the fact that the fake claim in each article was not actually supported by the fake evidence, in whatever form it was presented.

In the next experiment the researchers wanted to rule out the possibility that just any complex, scientific-looking image would make people think the article was more scientifically credible than it really was. This time they included a topographical map of brain activation, which does not look like a brain and rarely could be seen in the media. Once again, the article with the brain image was rated higher for scientific reasoning.

In the last experiment, McCabe and Castel modified a real write-up of a real brain-imaging study, which argued that brain imaging can be used as a lie detector. Students read one of two versions of the article. One of them contained criticism from a brain researcher who questioned whether the technique would work outside a laboratory, while the other omitted the criticism. The students were put into groups in a typical experimental design (two groups with the brain image present or absent, and two groups with criticism present or absent). The agreement with the article's conclusion was higher when a brain image was presented, even though the same evidence was presented in textual form in the article, making the brain image redundant. Even more striking, while agreement with the conclusion was affected by the presence of the brain image, the presence or absence of substantive criticism had no effect. Criticism did have an effect on whether the students thought the article title ("Brain Scans can Detect Criminals") was appropriate, however.[142]

How persuasive this brain porn can be in affecting consumers' mindsets and attitudes is perhaps well known to marketers, as shops are groaning with an array of products with images of brains on them, or at least with the miracle prefix "neuro" in their name.

It is not by coincidence that I began this chapter with a reference to the growing popularity of phrenology. There are more similarities than differences between phrenology and neuroscience. Phrenology attempted to localize cognition processes in the brain, and this is also neuroscience's area of interest. Phrenology promised to bring sensational results in diagnosis and neuroscience likewise has awakened similar expectations. But the most striking resemblance lies in the fact that marketing seems to be the driving force behind the public image of both fields.

"Wait a minute!" someone might reflect. "Marketing is one thing and science is another. Scientists can't be blamed for what's going on in the consumer market, which is driven by greed."

And yes, indeed these two things should not be considered as one. However, this book is about cargo cult in science, therefore we ought to look carefully into what makes neuroscience flourish in such plentiful and perverse marketing forms. Are scientists entirely innocent? Let us examine this problem more deeply. It might appear that no one but scientists overcome by their visions were the first to awaken unrealistic expectations and hopes. It might also appear that the market has only developed and made use of some of their ideas.

Franz Joseph Gall, an esteemed Austrian-German anatomist, worked on phrenology in the hope of transforming it into a complex science of brain function and human behavior. He believed that the mind in its entirety was situated within the brain. Favoring the theory that the analysis of the human skull could be a source of valuable knowledge of human nature, he made a lot of amazing claims and promises. During his triumphant European tour between 1805 and 1807, Gall lectured to crowned heads of state, universities, and scientific societies, raising the expectations of scientists as well as the public. These expectations never faded away

About ninety years later, when German physicist William Conrad Roentgen invented the X-ray technique, such hopes were revived. With this discovery, the ability to noninvasively observe the brain constituted a huge leap forward. Typical X-rays offered only a dim shadowgraph of the brain's structure, but even this clouded view was almost a miraculous development at the time. Scientists were overjoyed with that breakthrough, the excitement so far-reaching that a French neurologist working at the renowned Parisian hospital La Salpetrière, Hippolyte-Ferdinand Baraduc, even claimed that he could use X-rays to photograph his own thoughts. He called the pictures "psychoicons." The X-ray, of course, is mute when it comes to the brain, let alone the mind, because the rays cannot pass readily through the skull's thick walls. Such ideas may seem bizarre for us now, but at that time, it should be remembered, La Salpetrière psychologists and neurologists used photography as an instrument to record the symptoms of neuropsychopatic patients, hence they are entirely of scientific origin. For Baraduc, an ancient dream seemed to have come real – immaterial and subjective thoughts could now be observed objectively. Science was now just one step from exposing the most secret aspect of human nature. Unfortunately, Baraduc and his "psychoicons" were not accepted by scientists and instead of pursuing a scientific career he found recognition in occult circles.[143]

The X-ray technique fell flat when it came to fulfilling the expectations of scientists, but soon a new discovery grabbed their attention and dormant hopes were awakened. In the first decades of the 20th century some extraordinary discoveries were made, originally in Poland and Russia, that allowed researchers to record brain signals through an intact skull. In 1913, in his experiments on curarized dogs, physiologist Vladimir Pravdich-Neminsky reported recording minor transcranial voltages, in what came to be called an "electroencephalogram" (EEG). The next step was made in 1925 by Hans Berger of the University of Jena, who published the results of a study of signals obtained through the intact skull of a human being.[144] Again, hopes for a magical technique that would allow behavioral scientists to directly examine and measure the activity of the human brain were brought to life. The number of research studies using EEGs shot up in short order. Just as brain scans are today mandatory in any research into the human mind, EEG recordings were considered the gold standard for research and diagnosis in those days. In fact, the impact of the discovery of the EEG on neurology, psychiatry, and psychology has been profound but there are still fundamental barriers to understanding the human electroencephalogram, mainly because the EEG is a cumulative measure of all neuronal activity in the brain, and therefore details of interactions within the myriad of neurons in the network are very often indistinguishable.

In the meantime, some severe criticism suggesting that EEGs may not even measure neuronal activity appeared. John L. Kennedy, one of the critics and an experimental psychologist from Princeton University, created a physical model which produced rhythmic electrical changes resembling alpha waves in an electroencephalogram (EEG). The model was a gelatinous mass contained in a bowl and driven by a mechanical pulse. A bimetallic strip inside the bowl acted as a single dipole which electrochemically activated the gel. Electrodes on the outer surface of the bowl recorded impulses of electrical waves, whose frequency was similar to the normal alpha rhythm. Kennedy equated the activated gel with the brain, the mechanical pulse with the cerebrospinal fluid pulse, and the electrical output of the model with the alpha rhythm of the EEG. He proposed that alpha arises from mechanical driving of the brain and concluded that disruption of the anatomical coverings of the brain should change the alpha rhythm. Kennedy's model assumed that the brain is the electrical equivalent of a single dipole.[145]

However, the greatest disillusionment of the EEG enthusiasts was with its inability to provide tight and reliable correlations between the brain's functions and psychological states. When it eventually seemed that years of searching for a miracle tool were wasted, new ideas sprang up and old hopes were revived once again. The 20th century was a time of rapid proliferation of new technologies applied in the process of understanding the human brain and mind. This book does not aspire to offer detailed descriptions of them all, and therefore I will only mention a few. An inquiring reader will find many other good sources that elaborate their history and functioning.[146]

Neuroimaging owes its stunning career to two technologies. The first one, developed in the early 1970s, is computerized axial tomography (CT or CAT). It uses high-density X-rays to capture images in slices and produce a three-dimensional model of the brain's anatomy. A decade later the second one was devised, which became known as magnetic resonance imaging (MRI). It enables static problems, such as tumors, blood clots, and deformed blood vessels, to be detected. Both technologies are referred to as structural as they provide valuable information about fixed anatomy. Unfortunately, they both fail to offer knowledge that would throw light on the functioning of the brain.

The landscape changed in the 1980s with the emergence of positron emission tomography (PET), a three-dimensional functional imaging technique. It allows images of the brain in action to be taken. PET measures brain metabolism by deploying radioactive tracer molecules. The main assumption is that when brain cells are active, they need more energy in the form of glucose or oxygen.

A new technology, which has proved to be even more effective, came in the 1990s. Functional magnetic resonance imaging (fMRI) has become the most prominently used technique for measuring brain activity. The advantages of this technique lie in the higher spatial and temporal resolution that it offers, as well as in the fact that it does not involve radioactive material. fMRI is based on the observation that everything in the brain which enables us to feel, think, perceive, and act is correlated with higher oxygen consumption and regional blood flow in the brain. The relative concentrations of oxygenated or deoxygenated blood in a small area of brain tissue creates a signal known as the BOLD (blood-oxygen-level-dependent).

Both structural and functional techniques of brain imaging have many benefits and can help us understand much about the human brain. But at the same time they have many limitations. The cause for concern is that it is scientists, not the market, who have triggered unrealistic expectations, despite being fully aware of their inherent drawbacks.

In Chapter 4, specifically in the portion discussing the decade of the brain, I cited an enthusiastic view on neuroscience expressed by the editors of *Acta Psychologica*. This is not a unique opinion. Owen Jones, professor of law and biology at Vanderbilt, is even keener in his advocacy of the new field. In the middle of the first decade of the 21st century he joined a group of prominent neuroscientists and law professors who had applied for a large MacArthur Foundation grant.

> They hope to study a wide range of neurolaw questions, like: Do sexual offenders and violent teenagers show unusual patterns of brain activity? Is it possible to capture brain images of chronic neck pain when someone claims to have suf-

fered whiplash? In the meantime, Jones is turning Vanderbilt into a kind of Los Alamos for neurolaw. The university has just opened a $27 million neuroimaging center and has poached leading neuroscientists from around the world; soon, Jones hopes to enroll students in the nation's first program in law and neuroscience. "It's breathlessly exciting," he says. "This is the new frontier in law and science – we're peering into the black box to see how the brain is actually working, that hidden place in the dark quiet, where we have our private thoughts and private reactions – and the law will inevitably have to decide how to deal with this new technology."[147]

Does this not evoke the promises made by Baraduc and his "psychoicons" a century ago? Does it not remind you of prophecies formulated by cargo cult leaders? Certainly it is not the market, nor marketers, nor even journalists who have generated such far-reaching expectations for neuroscience in the field of law. But this is not the end of the story.

In 2006 the Society for Neuroscience published a press release entitled "New Research Gives Addicts Hope that Effects of Addiction Could be Reversed," in which they wrote:

> Recent findings show that the effects of addiction on the brain potentially could be reversed and indicate a clearer understanding of the powerful effects of brain circuits, self-control, and environment on drug taking.
> For years, addiction has been viewed simply as a weakness or defect in the addict, as reflected in the misguided "just say no" attitude. But gradually recent research has dispelled that myth, giving addicts and addiction the scientific and medical attention they deserve.
> "New avenues of research may provide new strategies for developing treatments and medications to treat cocaine and other drug abuse and dependence," says Michael Kuhar, PhD, at Emory University.[148]

I am writing this book nearly ten years later and I am sorry to say that the hopes awakened by scientists have not been fulfilled. As Sally Satel and Scott Lilienfeld write: "The clinical reality is just the opposite: The most effective interventions aim not at the brain but at the person. It's the minds of addicts that contain the stories of how addiction happens, why people continue to use drugs, and if they decide to stop, how they manage to do so. This deeply personal history can't be understood exclusively by inspecting neural circuity."[149]

I wish to see one day a neuroscientific cure-all for addictions, as it would mark a major step in improving human wellbeing. If there is even a small chance of finding such miracle solutions, we must not be prevented from exploring every single avenue that comes to light. And yet it is highly immoral to raise such unrealistic hopes, based on shaky foundations. Only one thing can come out of this, and not necessarily the most desirable: the weakening of the "just say no" attitude among addicts who try to cope by resisting temptations in this way. This once again proves that it is not the market that is responsible for this, but scientists themselves.

Neuromarketing is possibly the most rapid stream in today's flood of neurobabble. Again, when we look for its roots, we will find them at one of the best universities of the world. Gerald Zaltman, known as "the father of Neuromarketing," carried out his first experiments using PET scans in the Market lab at Harvard Business School in the 1980s. He and Stephen Kosslyn have even patented neuroimaging "as a means for validating whether a stimulus such as advertisement, communication, or product evokes a certain mental response such as emotion, preference, or memory, or to predict the consequences of the stimulus on later behavior such as consumption or purchasing."[150]

There are many, many more neuro(over)enthusiasts among scientists. At the most prestigious universities they have replaced traditional "experimental" or "cognitive" teaching programs with cognitive neuroscience programs. They infect others by organizing "Brain Carnivals"[151] and even "Brain Days" for children.[152] None of the foregoing should imply that science should not be popularized. It definitely should, and there is still much to be done in this respect. When doing so, however, we must avoid fueling unrealistic expectations, as this is the primary domain of pseudoscience.

Neuroscience has made strides in the last few decades, and neuroscientists are constantly discovering amazing things about the brain. But at the same time cargo cult-like thinking has contaminated this discipline and has caused unsubstantiated expectations and a myriad of groundless claims. Fortunately, there are people with calm and realistic minds who can dampen this immoderate fervor. One of them is Martha Farath from the University of Pennsylvania, who wrote one of the first critical studies about neurocognitive psychology, or more specifically about the locality assumption that frequently underlies the use of neuropsychological data in the development of cognitive theories. She argues that it is commonly used incorrectly. Her paper discusses the general implications of denying the locality assumption, and includes an examination of the modularity hypothesis and the choice between top-down and bottom-up research approaches.[153]

William Uttal, professor of psychology at Arizona State University, in his book entitled *The New Phrenology: The Limits of Localizing Cognitive Processes in the Brain*[154] also questions the concept of localizing mental processes by using innovative functional brain imaging techniques. Yet his aim is to criticize the revival of the localization concept rather than the neuroimaging technology itself. The excitement aroused by the new neuroscience "toys" that enable us to look deep inside the brain resembles a dog chasing its tail. Relevant new data appear almost daily.[155] The more data we get from research into localization of mental functions, the more doubts arise as to whether we have properly defined the subject of the research. Each and every publication seems to bring us closer to the belief of the 19th-century science that one can explore an individual's personality by simply palpating the lumps on their skull. It seems that the key difference is in our confidence that the contemporary "palpation" is much more precise and thus better somehow. And yet it is not possible to clearly localize the so-called higher mental functions, as they require activation of various parts of our nervous system. Rather than undermining neuroscience, Uttal trains his fire on the activities that I refer to as the cargo cult in this book.

The boundless enthusiasm of many neuroscientists must raise some suspicions. Such doubts were surely plaguing neuroscientist Craig Bennet when he went to the fish market to buy a dead (and frozen) Atlantic salmon to be used later in his research. Bennet and his colleagues placed their subject in a brain scanner, "showed" it pictures of people in various social situations and "asked" it questions about the feelings of people on pictures. During this experiment a small area in the salmon brain flared to life in response to the task. The researchers described their experiment and its results in an article entitled "Neural Correlates of Interspecies Prospective Taking in the Postmortem Atlantic Salmon: An Argument for Multiple Comparisons Correction."[156] The "salmon study" won a 2012 Ig Nobel Prize for work "that makes people laugh, then think." But the primary aim of Bennet's team was neither to win the Ig Nobel Prize, nor to demonstrate that the salmon's brain becomes activated while "watching" people in a variety of social situations. The team was composed of professional and experienced neuroscientists who realized that when researchers run large numbers of statistical tests simultaneously on BOLD signals, some of these tests are bound to appear as "statistically significant" simply as a result of chance. These results mistakenly show that the brain is more active while performing a task by the subject, when in fact this part of the brain is not working. That is why Bennet decided to show these artefacts in such an unorthodox way. Their study demonstrates that decisions taken in data analysis can severely impact the reliability of fMRI results.

Statistical artefacts are not the only weak link in the process of explaining human behavior that is offered by neuroscience. In 2008, an article was written by Edward Vul, a PhD student under the supervision of neuroscientist Nancy Kanwisher at the Massachusetts Institute of Technology in Cambridge, along with psychologists at the University of California in San Diego, including Harold Pashler. Even before its publication the article unleashed a storm among neuroscientists (formerly referred to as "experimental" or "cognitive" psychologists). The reasons were twofold: first, the authors showed that the king was naked by providing evidence corroborating that high correlations in fMRI studies of emotion, personality, and social cognition are close to worthless; second, the way the results of the analysis were presented sent scientists into a real frenzy. Having obtained acceptance for their publication, the authors decided to prerelease it on Pashler's blog and made it available to the public. The article's first title was "Voodoo Correlations in Social Neuroscience," which additionally touched a nerve.[157] Finally it was published as "Puzzlingly High Correlations in fMRI Studies of Emotion, Personality, and Social Cognition."[158]

The question that begs asking is: what was wrong with the correlations? The researchers selected 54 papers in the field of social neuroscience and sent a brief questionnaire to their authors requesting details of their analyses. Based on the collected information, they concluded that in a "red list" of 31 cases – often in high-profile journals, including *Nature* and *Science* – the authors made fundamental errors in data handling and statistics. The recurring problem in the studies under investigation was a *non-independence error*, in which the final measure (say, a correlation between behavior and brain activity in a certain region) is not independent of the selection criteria (how the researchers had chosen which brain region to study). This error, the researchers claimed, had its roots in selecting small areas of the brain, called voxels, on the basis of their high correlation with a psychological response, and then going on to report the magnitude of that correlation. In a nutshell, the trouble is that scientists search through a huge set of data for associations that are statistically significant to pinpoint those that are most promising. Further analysis is performed only on those associations which have been cherry-picked for scrutiny.

This is the major, although not the only, problem described by Vol and his colleagues. The remaining areas of concern are of a more technical nature. For those who wish to explore the subject in more detail, the article is still available online.[159]

What is worthy of comment is how particular scientists reacted to these assertions. It is always interesting to observe the behavior of people against whom accusations of misconduct or even dishonesty are brought publicly. In particular, this refers to those whose responsibility and life's dedication should be to search for truth.

As could be expected, some denied the accusations by explaining that there could have been no non-independent error because calculating the size is not a statistical test.[160] Others said they had not had the chance to argue their case in the normal academic channels. But still others, like neuroscientist Chris Frith of University College London, expressed their deep concern about the loss of reputation for their discipline: "We are not worried about our close colleagues, who will understand the arguments. We are worried that the whole enterprise of social neuroscience falls into disrepute."[161]

What a typical way to comport oneself in the face of criticism! Whenever I fault scientific practices publicly, almost always someone instantly returns fire by ascertaining that such issues should be discussed among academics, and not revealed for all to see and judge, because otherwise people's confidence in our discipline suffers. But why, as a matter of fact, should they not be exposed and subjected to the public gaze? As long as cargo cult practices are, without a hint of reflection, tolerated and cultivated within our field, the discipline deserves no trust from the public. Those of us who understand the weight of this dilemma have an obligation to warn society.

Among those condemned by Vul and his colleagues was, by a fortunate chance, one who saw the error of his ways and took a valuable lesson from the experience. It was Christian Keysers of the University of Groningen in the Netherlands who said: "On the other hand, we all agree that there is a kernel of truth in what Vul and his colleagues write about some of the literature being shaky. We can never be reminded often enough of the importance of good statistical practice."[162]

Another article takes us back to the circle of problems broached in the previous chapter. The raw material for this study was 49 meta-analyses, or studies that analyze data from other studies. In total, 730 neuroscience studies published in 2011 were covered. The team analyzing the data concluded that most of the reported findings may be of little or no reliability. The problem was low statistical power:

> A study with low statistical power has a reduced chance of detecting a true effect, but it is less well appreciated that low power also reduces the likelihood that a statistically significant result reflects a true effect. Here, we show that the average statistical power of studies in the neurosciences is very low. The consequences of this include overestimates of effect size and low reproducibility of results. There are also ethical dimensions to this problem, as unreliable research is inefficient and wasteful. Improving reproducibility in neuroscience is a key priority and requires attention to well-established but often ignored methodological principles.[163]

Alas, another analysis of the neuroscience studies carried out in 2013 returned a similar result. The authors reported that *prospective statistical power*, or at least the level of reference to statistical power, was observed in only 5% of all 183 empirical articles published in *Psychological Science* in 2012. The analyzed studies lacked sufficient power to detect anything other than large effects, around 20% of the articles. There is still poor reproducibility of scientific findings, as well as sub-par publication quality.[164]

One simply cannot escape the feeling that this is a new science, armed with extortionately expensive and state-of-the-art technologies and hailed as the hope of the social sciences, going down the same erroneous path.

Challenges facing cognitive neuroscience were well captured in the title of an article written by John Cacioppo and his coworkers: "Just Because You're Imaging the Brain Doesn't Mean you can Stop Using Your Head."[165] The authors outline a set of basic principles designed to help make sense of brain-imaging research within the fields of cognitive and social neuroscience. They begin with a principle few would debate – that social cognition, emotion, and behavior involve the brain – but whose implications might not be entirely obvious to those new to the field. Other principles call attention to points at issue still present in neuroscience:

Principle 2: The Functional Localization of Component Social Processes or Representations Is Not a Search for "Centers"

Principle 3: Localized Changes in Brain Activation That Differ as a Function of a Task Do Not, in Themselves, Signal a Neural Substrate

Principle 4: The Beauty of a Brain Image Does Not Speak to the Psychological Significance of the Image[166]

While putting the results of neuroscientific research under scrutiny we should bear in mind that the splashes of color that scientists see on their monitors is the concentration of oxygen dissolved in the blood. The relative concentrations of oxygenated blood in a small area of brain tissue creates a signal known as the BOLD response. There is still a huge gap between BOLD responses and specific human behavior that is full of enigmas yet to be deciphered deciphering. Some neuroscientists who have rhapsodized over the assumed progress of their discipline seem to be forgetful about the fact that, rather than focusing on the neural activities alone, it would be of greater value to get a better understanding of behavior patterns and cognition processes.

In 2008, Nikos Logothetis, director of the Max Plank Institute for Biological Cybernetics and one of the pioneers of fMRI, published a surprisingly harsh paper in *Nature* about the limits of fMRI as an experimental tool. The piece is especially noteworthy because Logothetis has probably done more than anyone else to document the tight correlation between what fMRI measures (changes in cortical blood flow) and the underlying neural activity of the brain. Logothetis hits the important points of what it is we think we are actually measuring using the fMRI.

> In humans, fMRI is used routinely not just to study sensory processing or control of action, but also to draw provocative conclusions about the neural mechanisms of cognitive capacities, ranging from recognition and memory to pondering ethical dilemmas. Its popular fascination is reflected in countless articles in the press speculating on potential applications, and seeming to indicate that with fMRI we can read minds better than direct tests of behaviour itself. Unsurprisingly, criticism has been just as vigorous, both among scientists and the public. In fact, fMRI is not and will never be a mind reader, as some of the proponents of decoding-based methods suggest, nor is it a worthless and non-informative "neophrenology" that is condemned to fail, as has been occasionally argued.[167]

Sally Satel and Scott Lilienfeld, in their book *Brainwashed: The Seductive Appeal of Mindless Neuroscience*, list six reasons why we should be skeptical about discoveries that kick off with the formula "Brain Scans Show…":

> First, brain scans rarely allow investigators to conclude that structure X "causes" function Y. …

Second, the subtraction technique used in most fMRI experiments is not necessarily well suited to the question being asked. ...
Third, although neuroimaging has deepened our knowledge of brain anatomy and function, its popular application tends to reinforce the misbegotten notion of the brain as a repository of discrete modules that control distinct capacities to think and feel. ...
A fourth caveat to keep in mind when one is interpreting brain scans is the importance of experimental design. The way in which investigators design their task can exert a big impact on the responses they obtain. ...
A fifth caveat stems from the fact that fMRI is an indirect method. Contrary to popular belief, imaging does not measure action of brain cells per se. ...
Last, it is important to keep in mind that before the final data even "reach" the voxel, analysts must deploy statistical approaches to extract meaningful information from noise.[168]

These restrictions demonstrate how the real picture of neuroscience is far from its colorful marketing image. As neuroscientist and psychiatrist Thomas Insel concluded in his article from 2009, there is no evidence that the past two decades of research in neuroscience have brought about a decrease in mental disorders, nor have they had any impact on their prevalence or on a patient's life expectancy.[169] There is also no specific evidence that neuromarketers can manipulate us using information gathered from scan images, in order to turn us into docile customers.[170] The expectations that neuroscience would help us to treat addictions have also not yet been fulfilled. Brain-based deception detection can perform impressively in laboratories, but there is no evidence that its capabilities can be used in natural forensic situations. There is a huge gulf between commercialized neuroscience and its actual power, which will probably never be spanned.

And last but not least, there are some problems which have not been mentioned yet. Neuroscience, for all its tools and toys, is not a game that scientists play in the solitude of their laboratories. The results of their poor research may have an impact on decisions to continue or terminate clinical trials with (oftentimes mentally or physically ill) humans as participants. Neurolaw enthusiasts could soon convince politicians to establish laws that will allow the use of brain scans as a proof of guilt in courts. The new brain sciences and technologies lead to people being categorized in new ways, and this, in turn, may adversely affect their social and personal life. We must bear in mind all of the possible social consequences of each decision made by researchers, regardless of the neuroimaging technology chosen, statistical method used, and, above all, particular interpretation of the results.

---

[137] C. Neider, ed. *The Autobiography of Mark Twain*. (New York: Harper, 1959): 69-73.
[138] Ibid.

[139] D. Carlat, "Brain Scans as Mind Readers? Don't Believe the Hype," *Wired Magazine*, *16.06*, (2008): http://archive.wired.com/medtech/health/magazine/16-06/mf_neurohacks?currentPage=all

[140] M. Crockett, "Beware neuro-bunk," *Ted conference*. (2012): http://www.ted.com/talks/molly_crockett_beware_neuro_bunk

[141] C. Chabris, and D. Simons, *The invisible Gorilla: And Other Ways our Intuitions Deceive us* (New York, NY: Crown Publishing, 2010): 139; P. Tracey, and D. Schluppeck, "Neuroentreprenuership: "Brain Pornography" or New Frontier in Entrepreneurship Research?" *Journal of Management Inquiry*, *23*, (2014): 101–103; http://www.sociallypsyched.org/item/brain-porn-neuro-bunk

[142] D. McCabe, and A. D. Castel, "Seeing is believing: The Effect of Brain Images on Judgments of Scientific Reasoning," *Cognition*, *107*, (2008): 343–352.

[143] E. Schenkel, and S. Welz, eds. *Magical Objects: Things and Beyond*. (Berlin: Galda and Wilch Verlag, 2007): 135-140.

[144] H. Berger, "Über das Elektrenkephalogramm des Menschen," *Archiv für Pssychiatrie*, *87*, (1929): 527-570.

[145] J. L. Kennedy, "A Possible Artifact in Electroencephalography," *Psychological Review*, *66*, (1959): 347-352

[146] For an overview of the history of neuroimaging, see B. H. Kevles, *Naked to the Bone: Medical Imaging in the Twentieth Century* (New Brunswick, NJ: Rutgers University Press, 1997); *Human Functional Brain Imaging, 1990-2009* (London: Wellcome Trust, 2011), http://www.wellcome.ac.uk/stellent/groups/corporatesite/@policy_communications/documents/web_document/WTVM052606.pdf; W. R. Uttal, *The New Phrenology: the Limits of Localising Processes in the Brain* (London: MIT Press, 2001).

[147] J. Rosen, "The Brain on the Stand," *The New York Time*, (March 11, 2007): http://query.nytimes.com/gst/fullpage.html?res=9A03E1DF1531F932A25750C0A9619C8B63

[148] http://www.sfn.org/Press-Room/News-Release-Archives/2006/NEW-RESEARCH-GIVES-ADDICTS?returnId={0C16364F-DB22-424A-849A-B7CF6FDCFE35}

[149] S. O. Lilienfeld, and S. Satel, *Brainwashed: The Seductive Appeal of Mindless Neuroscience* (New York: Basic Books, 2013), 70.

[150] U.S. Patent 6,099,319 was issued on August 8, 2000 to Gerald Zaltman and Stephen Kosslyn.

[151] http://news.wustl.edu/news/Pages/24453.aspx

[152] http://tuna.tulane.edu/Our_Events.html

[153] M. J. Farah, "Neuropsychological Inference with an Interactive Brain: A Critique of the 'Locality' Assumption," *Behavioral and Brain Sciences*, *17*, (1994): 43-104.

[154] Uttal, *The New Phrenology*.

[155] E.g., http://www.biopsychology.com

[156] C. M. Bennett, et al., "Neural Correlates of Interspecies Prospective Taking in the Postmortem Atlantic Salmon: An Argument for Multiple Comparisons Correction," *Journal of Serendipitous and Unexpected Results*, *1*, (2010): 1-5.

[157] J. Bardin, "The Voodoo That Scientists Do," *Seed*, (24 February, 2009): http://seedmagazine.com/content/article/that_voodoo_that_scientists_do/

[158] E. Vul, C. Harris, P. Winkielman and H. Pashler, "Puzzlingly High Correlations in fMRI Studies of Emotion, Personality, and Social Cognition. *Perspectives on psychological science 4*, (2008): 274-290.

[159] http://www.pashler.com/Articles/Vul_etal_2008inpress.pdf

[160] T. Singer, B. Seymour, J. P. O'Doherty, K. E. Stephan, R. J. Dolan, and C. D. Frith, "Empathic Neural Responses are Modulated by the Perceived Fairness of Others," *Nature, 439*, (January 2006): 466-469.

[161] A. Abbott, "Brain Imaging Studies Under Fire," *Nature, 457*, (January 2009).

[162] Ibid.

[163] K. S. Button, J. P. A. Ioannidis, C. Mokrysz, B. A. Nosek, J. Flint, E. S. J. Robinson, and M.s R. Munafò, "Power Failure: Why Small Sample Size Undermines the Reliability of Neuroscience," *Nature Reviews Neuroscience, 14*, (2013): 365-376.

[164] I. Vankov, J. Bowers, and M. R. Munafò, "On the Persistence of Low Power in Psychological Science," *Quarterly Journal of Experimental Psychology 67*, (2014): 1037-1040.

[165] J. T. Cacioppo, T. S. Lorig, G. G. Berntson, C. J. Norris, E. Rickett, and H. Nusbaum, "Just Because You're Imaging the Brain Doesn't Mean you can Stop Using your Head: A Primer and Set of First Principles," *Journal of Personality and Social Psychology, 85*, (2003): 650-661.

[166] Ibid.

[167] N. K. Logothetis, "What We Can Do and What We Cannot Do with fMRI," *Nature 453* (June 12, 2008): 869–78. http://kyb.mpg.de/fileadmin/user_upload/files/publications/attachments/NikosNatureJune2008_%5b0%5d.pdf

[168] Lilienfeld, Satel, *Brainwashed*, 14-19.

[169] T. R. Insel, "Translating Science into Opportunity," *Archives of General Psychiatry, 66*, (2009): 128-133.

[170] Lilienfeld, Satel, *Brainwashed*, 45.

Chapter 6: Why Has Science Transformed into a Cult?

My investigation of the accomplishments and challenges of the social sciences feels like a journey on a river that parts the territory of a most amazing land. The stretch on one bank encompasses a vast area of barren fields, occasionally overgrown with enormous masses of weeds. The people engage themselves in performing odd rituals, instead of cultivating their fields. By contrast, the land on the opposite bank has fertile soil on which fruitful plants are grown by determined and hardworking people. It has always been my belief that the social sciences hold magnificent and, at the same time, still underestimated, potential that can facilitate the growth of civilization as well as an improvement in human welfare. Consequently, I am confident that the land on both river banks is capable of yielding a similarly abundant crop. How can this be achieved? It cannot be done without first understanding why so many activities undertaken in the social sciences turn into a cargo cult in the first place.

The native inhabitants of the New Guinea mountains had no previous experiences to which they could compare their encounter with the "giant bird from heaven". The hypothesis stipulating that it must have been the spirits of their ancestors was entirely rational in the context of the knowledge they had established. Their minds, as ours would in a similar situation, sought to explain the unprecedented developments they had witnessed. In a similarly consistent manner, they were trying to put together pieces of a puzzle that made up a complicated picture, which we term a cargo cult.

Science is no different in this respect. Based on available knowledge and previous experiences, we formulate conceivable hypotheses. Yet sometimes their substantive worth is not much higher than the ones put forward by the New Guinea islanders. However, in order to realize that this problem is not exclusive to the social sciences only or even psychology in particular, it is worth recalling just a few of the most disconcerting examples in other fields of science. In those cases, the term cargo cult is perhaps even too gentle, especially given the terrifying implications of the concepts described.

In December 1949, Egas Moniz received the Nobel Prize in medicine from the Royal Swedish Academy of Sciences for "for his discovery of the therapeutic value of leucotomy in certain psychoses."[171] Leucotomy, also known as lobotomy, is a neurosurgical operation in which nerve fibers connecting the frontal lobes and the thalamic areas in the brain are cut. Early into the 20th century, this method was used in an attempt to heal schizophrenia and other serious mental disorders. One of its most fervent advocates, Walter Jackson Freeman, developed a prefrontal lobotomy procedure that involved inserting – through the eye socket – an ice pick to the patient's brain! He performed this type of surgery on more than 2,900 patients. By awarding the Nobel Prize for lobotomy, the world of science honored an invention that, in fact, was later proved to be the implementation of an erroneous hypothesis and often resulted in patients suffering excruciating mutilation. For many people, this hypothesis bore far worse consequences than the act of scattering roast pork over the fields as a sacrifice offered to gods. In the years 1935–1960 nearly 50,000 lobotomies were performed in the United States alone. Almost 60% of the patients died, some became mentally disabled, and only certain symptoms subsided in a few patients. Despite protests from one of the lobotomy victims, Christine Johnson, the Nobel Committee refused to revoke Moniz's award.

Another example is related to the origins of the beriberi disease. It took scientists several decades to understand that the hypothesis concerning its causes had been built on mistaken assumptions. In 1886, the Dutch government sent Christiaan Eijkman, a disciple of the famous microbiologist Robert Koch, to Western India to determine the causes of that disease. The approach to medicine in that era was thoroughly bacteriological. Eijkman spent many years trying to identify "a germ" spreading the disease. He was as strongly attached to the hypothesis as cargo cult followers were to the conviction that their forefathers sent them "great birds from heaven." Even though he came within a hair's breadth of finding the actual causes of the disease, neither he nor other investigators around him were able to discern them. It was only Frederick Hopkins who analyzed Eijkman's observations of hens suffering from beriberi. He detected that their condition might have been affected by the unpolished rice with which the animals had been fed. Hopkins demonstrated that it contained a substance that the human body was not capable of producing autonomously. Its *lack* was manifested by avitaminosis that with time developed into its extreme form, beriberi.

A similar and much more recent failure of science, mentioned in Chapter 2, represented an unfounded belief that key contributing factors behind gastric and duodenal ulcers were stress and unhealthy living habits. Although science has already rejected this explanation, on many occasions, particularly in articles whose aim is to advertise psychotherapy, one may still learn that psychology can cure also peptic ulceration disease.

Should therefore a conclusion be drawn that scientists make their hypothesis in the equally accidental manner in which cargo believers formulated their beliefs? No. It is true that both lines of reasoning are consonant with features of our mind and none of them entails the "hit-or-miss" method. The difference lies only in the pool of knowledge that is used by the scientific way of thinking. Science is much more scrupulous: in the process of science, we accumulate descriptions of occurrences, analyze them, and make deductions. Furthermore, scientific reasoning is characterized by a substantially higher number of *self-correcting* measures, which are much rarer in cargo cult practice (though not unheard of at all). In Chapter 1 I have quoted accounts showing how the cargo cult followers improved their "radio shacks" by, for instance, replacing "insulators" and "cables." These were truly self-correcting measures

– analogous to the ones performed by scientists themselves! Scientists also frequently try to improve, enhance, or modify things in experiments they conduct. A systematic approach to observation and action is perhaps yet another differentiating factor in this respect. A cargo cult believer with scientific ambition should not replace all "insulators" and "cables" at one time. They should do it one by one in order to be able to verify the outcome, and they should keep a register of their observations. Should all these measures be ineffective, they could also venture into questioning their own hypotheses and re-embark on observation of the white men, if they assume that this is the stage at which a mistake must have been made.

The cargo cult worshippers were not and are not foolish or primitive people. In many cases they reached highly or even extremely rational conclusions in the context of knowledge they had developed. The passage below reveals the manner in which the Upikno cult community modified their beliefs:

> When, however, people became bored or frustrated with the protracted waiting for cargo that had continuously failed to come, they decided to revise the prophet's teachings; it was decided, however, that traditional living habits should be restored. Therefore, it was necessary for everybody to burn European clothing and start wearing traditional jewelry and clothes. People returned to gardening routines and cultivation of edible plants. Prayers and ceremonies adopted from missionaries were not abandoned as it was agreed that traditional rituals and sacrifices, combined with Christian prayers and rites, would be more effective in propitiating the ancestors to ensure the arrival of cargo. Nevertheless, when after nearly five years of uninterrupted waiting, the cargo ship still did not arrive, people started losing trust in the Upikno prophet. Rituals worshiping deities gradually vanished, along with the cult itself.[172]

Can there be a more reasonable manner of verifying one's own judgments? If all followers had adopted a similar course of action, cargo cults would most likely have been long forgotten. In fact, some of them have indeed disappeared, while others were reborn in another form and yet others continue to prosper unchanged. The quoted account proves, however, that the natives did possess sufficient knowledge and intellect to be able to discard erroneous views in the same way as scientists who, having formulated mistaken hypotheses, after many years eventually discover their mistakes (or someone else does it for them). Hence there are no fundamental differences between "our" and "their" reasoning. And still, the cargo cult phenomenon is virtually absent in physics, chemistry, and mathematics, yet it runs rampant in both psychology and sociology. Let us therefore have a closer look at why some cargo cults flourish for such a long time – both on the Pacific islands and in the social sciences.

The first and foremost reason for the persistence of a cargo cult, be it among the inhabitants of the Pacific Isles, scientists, or social sciences practitioners, is simply *fraud*.

> In early 1964, Weworuya started calling upon the local people to tuck away their money in small, red wooden cases (boxes) and to deposit them with him. To that end, he erected a special, large storehouse and piled there an ever increasing number of boxes full of cash brought in by people from the Kanite vicinity. He claimed that after some time, when the cases were opened, they would be filled with money. However, for this to happen, rituals in honor of ancestral spirits should be performed, sacrifices to them should be offered and intensive prayers should be said. To demonstrate the credibility of his teaching, he took out money from all the cases, an already considerable amount, and travelled with it from village to village to show it to people, explaining that it was already the first fruit of his labor: the money multiplied. In every village he visited, he held night meetings and performed rituals in honor of ancestors and encouraged people to deposit more and more money since they would receive even more in return.[173]

One cannot tell for how long that prophet's lies could have remained undiscovered. Administration officials stepped in and prohibited him from continuing his fraudulent activity. However, we can rest assured that it was a very powerful manipulation, since Weworuya's habit of depositing money in wooden cases for subsequent multiplication endured and re-emerged in other cults as well.

Regrettably, when confronted with a deliberate deception, we are usually helpless. The methods used by Weworuya to manipulate New Guineans are still successfully implemented by others, for example in pyramid schemes. An articulate "Weworuya manager" can with great self-confidence teach others how to multiply money. His followers include many educated and (seemingly) reasonable people. Both priests and financial pyramid gurus feed on the human desire for things that people lack. How does this relate to science? The cargo cult-type scientific practices too often take advantage of a craving for what is unreal and what we can see and simultaneously envy in others. Common sense would indicate that the New Guinea islanders should be warned that no matter how hard they worked, they would only be able to acquire at best an insignificant chunk of the goods the white men had. And no rituals would help them in getting it. At the same time, science confronts people diagnosed with cancer, autism, or Down syndrome with the same severe prospects. Here again, those unfortunate patients will not change or cure certain symptoms, regardless of how hard they themselves or their parents try. Can anyone be surprised at the seductive power of prophecies that things could really be different, and how supporting such prophesies with scientific findings (ultimately as authentic as the suitcase of money collected by Weworuya) captures the naive and uniformed?

In my earlier book, *Psychology Gone Wrong: The Dark Sides of Science and Therapy*, we wrote about deliberate frauds where scientists falsify study findings, as well as about the consequences of such frauds. This book, as well as being a collection of previously described scientific misconduct, gives numerous descriptions of therapies whose structure is deceivingly similar to a cargo cult with, for example, unfounded news about miraculous recoveries, accusations of insufficient diligence in following the instructions of "prophets," or "scientific evidence" that has been ineptly forged – and all this against the backdrop of despair and hope.

Contrary to cargo cults, science has invented fairly effective tools for exposing fraudsters, such as controlled experiments or replications of research. Sadly, they remain commonly underused or they are employed in a way that leads to profound distortions in scientific evidence. What is worse, many "scientists" commit themselves to validating the efficacy of therapies (cargo) without having analyzed the empirical evidence first. This practice is well exemplified by the method originated by O. Carl Simonton and described in Chapter 8 in this book, or seen in therapies for the so-called adult children of alcoholics groups (the elucidation of which is to be found in Chapter 7). Many scientists, although aware of existing frauds, remain rather indifferent to the practices of cargo prophets that spread within their own professional fields. This indefensible sin of abandonment is committed more frequently by representatives of the social sciences than the natural ones.

A far more difficult phenomenon to analyze is *self-deception*. Here, it is the very process of rational inference that falls victim to considerable deformation. Consider how this mechanism keeps the faith in rituals alive, based on the "ancestral breathing cult":

> People partaking in those rites claimed to have felt a breeze of cold wind. And as it was locally believed that deities and ancestors wander from one place to another with the wind, in particular by night and with a cold wind, ritual trances were interpreted as effects of the breaths of passing deities and ancestors who were bringing cargo. Thus, in the belief of the Islanders, their ancestors were present, and if there was no cargo it only meant that the rituals were only partially effective; consequently, they should be enriched and complemented with new elements. Therefore, various stones, everyday goods, branches, shells, and plants were carried into the newly built cargo "houses," with a hope for the ancestors present during ritual trances to turn them into genuine European cargo.[174]

In this story, knowledge handed over from generation to generation on how spirits moved from one place to another became indisputable over time. It served as a reference for people in acknowledging the effectiveness of the ritual components. Hence, the cause of cargo failure must have lain in other stages of the ritual. Consequently, it was necessary to identify the flaw, refine the ritual, and anticipate the outcome.

The very same line of reasoning is followed by adherents of the Sigmund Freud cult and related religions.[175] Knowledge about resistance displayed by patients has been cultivated by generations since the times of the first "prophet", and the passage of time has rendered it unquestionable. During the rite of therapy, a patient's negative response is seen as undeniable proof of resistance (just as the blow of a cold wind was confirmation of the presence of the spirits). Therefore, the diagnosis must have been accurate and thus it validates "prophesies" made by psychoanalysis. What needs to be done then is to check what falls flat in other elements of the therapy, improve it, and continue with the ritual ceremonies.

Alas, self-deception is, by its very nature, non-collapsible. If we find no possibility to challenge truths revealed and preserved in tradition, we will be unable to break through the curtain separating cult from rational reasoning. In practice, all cults persist precisely where elements acknowledged by the tradition and experience of many generations are particularly multitudinous. Where subsequent "prophets" had created their system of prophecies from scratch and with no robust bonds with the traditional knowledge, cargo cults died out. It is surprising how abundant such cases were. In his superb monograph *Kulty Cargo na Nowej Gwinei* (*Cargo Cults on New Guinea*) that I have frequently quoted in this book, Władysław Kowalak gives a detailed depiction of 55 different cults. In as many as 18 cases the author concludes with the sentence "the cult has ceased to exist over time." Some cults have been terminated by formal administrative intervention and others absorbed by other cults, while we still lack sufficient data about some cults; however, there are some which continue to exist. In those 18 cases, the natives acted similarly to the followers of the Upikno cult referred to at the beginning of this chapter – that is, rationally. This can be ascribed mainly to the fact that their "prophets" had failed to forge sufficiently strong ties between their prophecies and the time-honored system of wisdom and beliefs.

An objective comparison with trends and schools popular in psychology and psychiatry is obviously difficult to find, therefore all of my comments below should be seen rather as a subjective reflection. If we assume that investigators of cargo cults explored all of their possible variations, we can conclude that there were only several dozen cults, at most no more than a hundred.[176] Meanwhile, within just one field of applied psychology – psychotherapy – there are several hundred different schools and modalities. As has been noted in *Psychology Gone Wrong*, in 1959 Robert A. Harper identified 36 various therapies.[177] By the end of the 1970s, Herink reported that the number of name-brand psychotherapies exceeded 250.[178] But as early as 1986 Daniel Goleman mentioned more than 460 various kinds of psychotherapy in his "guidebook,"[179] and by the turn of the 21st century revised estimates reached 500.[180] Nowadays they are certainly much more numerous, and attempts to count them all would be in vain. What is far more significant is that too few of them in fact disappear. Instead, they grow into other, allegedly more mature forms, or are modified and supplemented to surface under new brands. In my view, this is possible due to the fact they have successfully interwoven their own message and knowledge derived from the pseudoscientific tradition.

This mechanism also applies outside the field of psychotherapies. While writing about research on behavior, I mentioned the proliferative character of our discipline. Psychologists produce a raft of theories and theoretical constructs that are held to be valid for many years and are seldom or sometimes never put in a critical context. Can they all be true and useful?

A group of cults functioning within the mainstream of the social sciences or on the fringes of the pure sciences is characterized by peculiarly high potential for growth. This is because their prophets integrate elements of traditional wisdom with their own prophecies.

Neuro-linguistic programming (NLP)[181], or the oldest pseudo-science cultivated in the field of psychology – psychoanalysis, could easily belong to this category. The latter example is particularly significant given how long it has been present in the public imagination. Time adds weight to ideas. Those that have survived gain, at least in the general perception, more value only because they naturally became a part of the tradition. This is also a contributing factor behind the vigorous popularity of astrology and a whole host of other equally worthless theories that are the offspring of some ancient folk beliefs. Limited minds invest them with a specific value merely because they were formulated by some Chinese wise man 4,000 years ago and, to make things worse, they were written in incomprehensible, and thus even more mysterious, characters. The truth is that, unlike a bronze figure whose worth on the antique market has been boosted by a few centuries of being buried in the sand, an idea does not grow into something more valuable with time. In fact, it is just the contrary – time often ruthlessly devalues it.

Deliberate deception and self-deception represent two crucial mechanisms that keep cargo cults alive among both the inhabitants of the Pacific islands and social sciences practitioners. The third mechanism, hugely important and perhaps far more dangerous than self-deception, uses *confusion of reality and beliefs*. This is a bond that endows believers with long-lasting confidence in the truth of key underpinnings of their faith. Let us look at this mechanism based on the traditional example of the Pacific islanders to compare it to cargo cult practices in psychology.

During the Second World War, there were occurrences where – particularly in conditions of poor visibility – freight plane pilots would spot "air strips" and try to land on them, even in the middle of the jungle, most frequently with tragic results. Invariably, the side effect of such occurrences was the delivery of some kind of cargo to those who had been waiting for, praying for, and exhorting the gods to send it. Can one imagine more conclusive evidence of the veracity of the entire belief system? Additionally, during air battles when a plane went down in the jungle, local inhabitants took it as a godsend. What came as a particular shock to cargo cult followers was the fact that among American soldiers located in military bases were people with a complexion as dark as their own. The natives saw this fact as an irrefutable proof that the ancestral spirits rewarded some of them and sent cargo. All such experiences strongly reinforced the faith in the coming of cargo and created favorable conditions for the cult to thrive.

A very similar pattern can be observed during psychotherapy every now and then. This will be thoroughly examined in the section dedicated to the Doman-Delacato method in Chapter 11. Obviously, a promise to solve all complications related to the mental development of children, which in fact has the value of a cargo prophecy, will sometimes come true. Indeed, in a high number of cases, there will even be some in which the diagnosis was inaccurate or too pessimistic. All of a sudden, a miraculous healing takes place and, consequently, the followers' faith intensifies. The necessity to devote themselves passionately to rituals and therapeutic incantations should now be emphasized more strongly because the therapy has proven to be successful! Likewise, in patients suffering from tumors, a certain rate of unexplained, spontaneous remission is observed. The tumor cells stop multiplying and the patient recovers from the disease. If a clever cargo psychotherapist accompanies such a patient throughout the process, continuously making him believe that positive thinking may have healing powers and engaging the patient in related rituals, this fortuitous turn of events is hailed as a miracle that confirms the credibility of prophecies.

Some pseudoscience systems bearing the characteristics of a cargo cult have from day one incorporated elements derived from true science that are long known to have worked and given positive results. One of many examples that could again be cited here is neuro-linguistic programming, which unscrupulously steals some achievements of psychology in order to subsequently sell them as "NLP system-specific techniques." Principles of classical conditioning – even if verbalized with the jargon of the trend and applied by NLP therapists and trainers – work precisely as described by behaviorists. The presence of such methods, combined with a full arsenal of ineffective and often damaging practices, helps to maintain an impression that other techniques on offer can also yield expected results.

Such examples of confusing reality and beliefs are myriad. Numerous therapists eagerly benefit from the effect of the passage of time and expect to be credited with any changes that occur – including ones that emerge naturally over time. Natural processes that take place in the course of mental development and also observed in mentally handicapped children are used in the same way. Regardless of what type of therapy is implemented, be it canine-assisted therapy, educational kinesiology or other cargo-logy, children will develop spontaneously anyway. What is indeed the real talent of a cargo priest is the ability to take credit on account of this. While the New Guinea islanders had in fact no frame of reference to verify whether the appearance of a plane on their air fields was the result of chance or of their prayers, we do have a variety of such tools at our disposal. As mentioned before, they include randomized controlled trials (RCT), double-blind experiments, or well-constructed statistical analyses to differentiate between coincidence and regularity. If we wish to do so, we have every possibility to use them (without limiting ourselves to self-reports only) to study behavior. If we reject them, this is typically done for one of three reasons: foolishness, ignorance, or protection of one's own dishonest interests.

The fourth, substantial reason underlying the persistence and revival of cargo cults is *deliberate abandonment and discouragement of effective evaluation*. Empirical evidence is a deadly threat to all cults. Canny leaders of cargo cults are well aware of it, as are a group of social sciences practitioners who are deeply engaged in cargo cult–type activities. Those who have ignored or ignore this source of knowledge condemn themselves to non-existence. The following accounts show the twilight of cults and prophets that comes from their world clashing with empirical cognition:

> Timo attracted many followers, and the cult itself was very vibrant, yet equally short-lived. Previous prophecies by Timo had not come true, and people were experiencing a time of disillusionment and spiritual dissatisfaction. …
> The revelations, however, did not come to pass, and the cult started to fade away. …
> When, however, the flood had not come and the prophecies had not been fulfilled, the frustrated people returned to their villages. Polelesi was left all alone with her prophecies, while the cult rituals were gradually dying out. …
> The cult gained numerous followers, who nevertheless started to leave their prophet with the passing of time because his prophecies had not come true.[182]

Among cargo prophets there were also those who were perfectly aware that if they had actually awaited the empirical validation of their own prophecies, they would have doomed themselves to this kind of sad ending. One of them, a certain Gogol, foretold of a great flood that would occur on Christmas (he even commanded people to build an ark) followed by the extermination of the white men. Then, the much awaited cargo would eventually arrive. "On Christmas Day, the 25th of December 1950, Gogol's wife Nene, while sleeping was given new instructions that consequently changed the entire teachings passed by Gogol up to that day. Nene claimed that in her sleep the messiah told her that they should all together leave 'the Noah's ark,' take all their belongings and rush away to the bush. Everything should be removed and cleared, otherwise all those present in the 'ark' would be beheaded."[183]

As it is easy to guess, no native, upon hearing Gogol's new message, cared to see the prophecy validated on empirical grounds. Obedient to the instruction, the believers destroyed the ark and retreated to the bush accordingly. The prophet enjoyed enduring respect for many years afterwards.

Adherents of pseudoscience are equally eager to emulate this pattern of behavior. In their version there can be no prophecies or changes. They rather seek to show that what is put to empirical verification and criticism has already passed into history, and the current model represents a totally different concept. As such, it naturally deviates from the one that inspired it, and therefore neither verification nor criticism can be applied in its case. This line of argumentation resembles the tricks played by the Greek god Proteus. He had the gift of fortune telling and could also morph his body into a variety of shapes. Anyone who wanted Proteus to reveal their fate had to capture and hold on to the god first. As he would turn into different animals or take the form of fire, a stream, or a stone, catching this slippery master bordered on the miraculous. It is equally difficult to formulate criticism of pseudoscience.

I have experienced this arduous labor myself when identifying the substantial flaws of neuro-linguistic programming. After I had demonstrated research findings that challenged the underpinnings of theses made by John Grinder and Richard Bandler, my opponents promptly enlightened me that I had overestimated the significance and contribution of both gentlemen in the formation of what NLP was today. When I raised ethical concerns related to tenets of the NLP model and practices, the response was that such doubts should be attributed to "generated and vulgarized NLP" only because "professional NLP" was "completely different." An attempt to challenge claims put forward in any of the books dedicated to NLP is countered with an argument that this particular piece of writing has nothing in common with the NLP mainstream.

Criticism of psychoanalysis evokes a strikingly similar reaction. If strictures are made upon Freud, one might hear in return that psychoanalysis has drifted far away from its sources and has taken on a totally new identity. Criticism directed at some more modern trends in psychoanalysis will elicit a response that it is one of the less prominent directions, and thus the story goes on a *d infinitum*. Does this not bring to mind Gogol's prophecies? Build the ark – the flood is coming! Destroy the ark and leave it immediately, otherwise you will die! Each of these appeals sounds sufficiently threatening and their content changes quickly enough to ensure that they are unverifiable. Besides, even if one really felt the need to verify them after all, it would take long enough for Gogol and his followers to be able to hide deep in the forest.

Representatives of scientific cargo cults, however, not only make their arguments vague or incomprehensible so that it becomes virtually impossible to verify the authenticity of their claims and efficacy of their methods. Many actively discourage others from research and any form of validation of the authenticity of their claims. Consider the very example of Sigmund Freud. In 1934, in response to a letter sent by American psychologist Saul Rosenzweig, who suggested putting Freud's theses to empirical investigation, Freud concluded: "the wealth of reliable observations on which these assertions rest makes them independent of experimental verification."[184] This is not the only instance of Freud's distaste for empirical cognition. Rosenzweig throws more light on the famous neurologist's reluctance to have his theories tested: "On two separate occasions (1934 and 1937), first in gothic script and then in English, Freud made a similar negative response to any attempts to explore psychoanalytic theory by laboratory methods. This exchange clearly underscored Freud's distrust of, if not opposition to, experimental approaches to the validation of his clinically derived concepts. Freud consistently believed that the clinical validation of his theories, which were based originally and continuously on his self-analysis, left little to be desired from other sources of support."[185]

In turn, Glenn Doman, the originator of a well-known child therapy that is explored later in this book, gives the following guidance to parents: "If, after all this time and this wealth of experience, a parent should ask for some important advice in a very short sentence, that advice would be: *Go joyously, go like a wind and don't test.*"[186]

Freud was not alone in his endeavor to discourage scientific examination of therapy efficacy. Bandler, Hellinger, and many other cargo prophets acted in the same manner, as has been extensively described in *Psychology Gone Wrong*. Many therapists claim that the efficiency of psychotherapy cannot be measured because the process is too complex to investigate it. Founding fathers of therapies that will be addressed elsewhere in this book keep scientists away from their institutions and clinics, too. All these people were and are aware that too many controlled experiments and an empirically-based attempt of validation could put an end to their careers. That is why some of them fabricate research of dubious quality and formulate conclusions that are later used as slogans advertising cargo consultation rooms. As much as this approach, although reprehensible, can be understood as an urge to protect one's own interests, the attitude of scientists who support such activity is a thought-provoking phenomenon. It will receive a fair amount of my critical attention in the following part of this book.[187]

---

[171] http://www.nobelprize.org/nobel_prizes/medicine/laureates/1949/
[172] Kowalak, *Kulty Cargo*, 158.
[173] Ibid., 219.
[174] Ibid., 172.
[175] For a detailed description of psychoanalysis as a pseudoscience see: Witkowski and Zatonski, *Psychology Gone Wrong*.
[176] Dawkins, *The God Delusion*, 236.
[177] R. A. Harper, *Psychoanalysis and Psychotherapy: Thirty-six Systems* (Englewood Cliffs NJ: Prentice-Hall, 1959).
[178] R. Herink, ed., *The psychotherapy Handbook. The A to Z Guide to More than 250 Different Therapies in Use Today* (New York: Meridian Books, 1981).
[179] G. Goleman, "Psychiatry: Guide to Therapy is Fiercely Opposed," *New York Times*, (September 23, 1986).

[180] D. A. Eisner, *The Death of Psychotherapy: From Freud to Alien Abductions* (Westport, Conn.: Praeger, 2000).

[181] For a detailed description of NLP as a pseudoscience and a form of psychobusiness see: Witkowski and Zatonski, *Psychology Gone Wrong*.

[182] Kowalak, *Kulty Cargo,* 150, 181, 208.

[183] Ibid, 212.

[184] S. Rosenzweig, "Letters by Freud on Experimental Psychodynamics," *American Psychologist 52*, (1997): 571.

[185] Ibid.

[186] G. Doman, *How to Teach your Baby to Read. The Gentle Revolution* (Pennsylvania: The Better Baby Press, 1990), *ix*.

[187] The problems resulting from scientists' passive and active support for the world of psycho-business are discussed in: Witkowski and Zatonski, *Psychology Gone Wrong*.

PART II:
UNCONTROLLED EXPERIMENTS ON HUMANS: CARGO CULT IN PSYCHOTHERAPY OF ADULTS

> *Woe to the foolish prophets who follow their own spirit and have seen nothing!*
> —Ezekiel 13:3

Your perfectly planned and long awaited holiday amidst exotic scenery, as if taken from your favorite childhood books ... wrecked by a nasty parasite or other similar bugs. Who of us has not experienced a similar situation? Expensive drugs that we have prudently packed in the first-aid kit can at best hold the germs off for a while before they send us into a state of utter humiliation. Huddled on the seat of a car or coach and dazed with fever, we record in our memory scenes flashing by the window with at most the clarity of a TV program viewed when cleaning a room. And in our mind we go over every meal in a vain search for the mistake we have made, or that tiny drop of unboiled water which, like a Trojan horse, has forced its way into our digestive system. Never again!

Never ... and still, we do know that our immune system is not resistant to exotic parasite attacks, just as the immune system of native Americans was not suited to fight off smallpox, measles, influenza, diphtheria, typhus or other diseases brought into America by Columbus. There seems to be only one way to guarantee that such an adventure will not affect us again – staying at home. Scientists have even forged a name for a range of similar behaviors that are supposed to protect us from such infections, and that is *behavioral immune systems*. One of the most important functions of the behavioral immune system may actually be to make people avoid members of other groups – in other words, to close themselves off to outsiders. Corey L. Fincher and Randy Thornhill, both biologists at the University of New Mexico, describe three specific attitudes or behaviors that are associated with this group-protective aspect of the behavioral immune system. The first is philopatry, or the tendency to stay close to where one was born. Obviously, never leaving one's home village or territory ought to significantly limit one's exposure to strange bugs and viruses. The second is ethnocentrism, or in-group favoritism – the preference for socializing and living among people of one's own kind. And the third is xenophobia, or the active fear and dislike of outsiders.[188]

But these are not all. Fincher and Thornhill demonstrated that the magnitude with which people value strong family ties or heightened religiosity is a result of a contingent psychological adaptation that facilitates in-group assortative sociality in the face of high levels of parasite-stress. To put it simply, religion protects us from parasites. Moreover, their studies showed that countries with higher levels of different infectious diseases had more religious and more family-oriented citizens than countries with fewer diseases. Within the United States, a similar dynamic was discovered: states with high levels of transmissible illness reported greater religiosity and more preference for socializing with family instead of with outsiders than states with relatively less parasitic danger.[189]

This very interesting theory of religion differentiation helps us to understand why we have thousands of religions[190] and yet new cults continue to come onto the scene. In just one of the monographs that I have repeatedly quoted throughout this book, fifty-five cargo cult forms were identified. If they all play an evolutionarily adaptive role of protecting their believers from infectious diseases spread by strangers, that is to say followers of other religions, there seems to be a clear and understandable explanation for their number. However, if we assumed, just for a fleeting moment, that the followers of all those cults spend most of their energy on finding the true, most powerful religion with the most supreme of all gods at its center, surely such high numbers would refute that assumption. At the end of the day, had such a religion existed and had its disciples been truly engaged in bringing it to light, the numbers should have been declining, not increasing.

What I have attempted to show in the preceding chapters is that psychology, with its ever-expanding variety of theories, constructs and hypotheses, has become a field whose representatives claim to be hunting for the one and only, the scientific Holy Grail with the highest predictive value. Practitioners of applied psychology, that is therapists and scientists who have put therapeutic processes under scrutiny, comfort us that their goal is to find the most effective treatment methods. How is it therefore possible that these useless tools, instead of being abandoned and their impact reduced, remain constantly on the rise? Or perhaps, just as in religion, new approaches serve the purpose of isolating people from other groups of worshippers? Or, through diversification and polarization, they shield them from the parasites of doubt and viruses of critical thinking?

There is probably not a single area of psychology that has not been affected by cargo cults in one way or another. Nevertheless, whereas in science critical works trigger a fair amount of discussions or, in the worst case, are skipped over in silence and indifference, in psychotherapy any criticism instantly comes under fierce fire. This is what happened after the publication of the article *Current Status and Future Prospects of Clinical Psychology* by Timothy Baker and his colleagues,[191] and its subsequent brief discussion in *Newsweek*.[192] The work of Baker and his colleagues mercilessly exposed the ignorance of psychotherapists in the field of evidence-based psychotherapy (EBP) research. It showed that millions of patients who could have enjoyed the benefits of experimentally confirmed therapy were instead receiving, as ordered by cargo cult therapists, meditation, facilitated communication, dolphin-assisted therapy or other treatment procedures with more than a thousand variants, most of which had never been seriously studied. The authors of the analysis demonstrated that American clinical psychologists put more trust in their own clinical experience supported by reports of their colleagues than in the scientific evidence. Many of them are even unaware that a concept called EBP exists and that it can be successfully implemented into therapeutic practice. The ultimate, and inevitable, conclusion is that clinical psychologists in the United States are practicing their profession in much the same way as they did in 1948, when the American Psychological Association (APA) accredited the first twenty-nine study programs in clinical psychology.

Baker and his colleagues suggest a significant change should be considered in the existing accreditation system within the APA. Among the many propositions they put forward, one aspect is particularly striking. They believe that the new system of accreditation should stigmatize unscientific training programs, as well as those practitioners who use methods that are scientifically unproven. They justify this call with similar changes that have been successfully put into practice in medicine.

The issued raised by Baker and his associates are not, however, exactly new. A few years ago, William O'Donohue and Kyle E. Ferguson, in an article called "Evidence-Based Practice in Psychology and Behavior Analysis,"[193] suggested that all therapies which are not empirically confirmed should be classified as an "experimental therapy," and that patients should be recognized as "medical subjects." They referred to drug testing procedures and proposed to give such patients the same rights as those who take part in new drug testing programs. It was also suggested that using problematic therapeutic interventions such as the "facilitated communication method" in the case of autistic children should be treated as a criminal act.[194]

Analogically, it should be further understood that the abandonment of drug testing procedures in favor of the application of untested therapeutic methods is a crime as well. One example of such malpractice could be analytical psychotherapy, used as treatment for mood disorders caused by abnormal functioning of the thyroid. I have met a few people who have experienced such an approach, which in several cases ended with tragic results. One of those stories is mentioned on my online blog:

> I was also a patient in the Neuroses Treatment Facility, where I was under the care of a physician and a psychotherapist (dynamic psychotherapy). The psychotherapist forbade me to take the prescribed drugs and the physician agreed. I (gradually) became extremely depressed, deconcentrated, scared and my mind was full of suicidal thoughts. I was unable to function properly both at home and at work. I was forced to resign from my job, while the crisis was getting even more severe, and finding a new one would be extremely difficult. I decided to stop therapy and started to take drugs again and with time my health

got back to normal eventually. I have been striving hard to this day, however, to regain my lost professional position and income.[195]

I do not fully agree with O'Donohue and Ferguson. I think that the term "experimental therapy" is insufficient or even misleading. It may imply that the therapist conducting the treatment skillfully controls its effectiveness by making relevant measurements throughout the entire process, comparing therapeutic groups with control groups, as well as by taking every measure that is necessary for determining the effectiveness of therapy. Unfortunately, a large number of therapists fail to meet such requirements or they even discourage others from doing so. A more precise term should be used to illuminate what these practices in fact are. I would propose the term "therapies in the phase of uncontrolled experiments on humans" as an alternative. It more closely reflects the reality, though it still seems to be far too moderate to define practices carried out by such authors (cargo cult priests).

Whether a given therapeutic procedure deserves to be called psychotherapy or should be red flagged and labeled as a therapy in the phase of uncontrolled experiments on humans is determined by standards such as the one applied by the American Psychological Association.

Well-Established Treatments. Well-established treatments meet the following criteria (I or II, and III, IV, and V):
I. At least two good between group design experiments demonstrating efficacy in one of two ways:
    A. The treatment in question is statistically significant to a placebo condition (pill or psychological) or to another treatment.
    B. The treatment in question is equivalent to an already established treatment in experiments with adequate sample sizes.
OR
II. More than a series of 9 single case experiments demonstrating efficacy. These experiments must employ a good experimental design and compare the intervention in question to another placebo or treatment, as is the case in I.A.
III. Experiments must utilize treatment manuals.
IV. Participant characteristics must be explicitly stated.
V. Results must come from at least two different research teams.

Probably efficacious treatments. Probably efficacious treatments share criteria IV and V though differ with respect to criteria I through III.
I. Two experiments that demonstrate that the treatment in question is statistically significant to a waiting-list control group.
OR
II. One or more experiments meeting all the criteria save criterion V.
III. Three or fewer single case experiments otherwise meeting all the criteria for well-established treatments.[196]

The time has come for the above specification to be updated. Just because therapy has proven effective in treating one condition, for instance phobias, it should not be automatically accepted as effective to treat another condition, for example depression. To determine whether it can be really put into practice, a separate evaluation process for each condition should be undertaken. Unfortunately, most of the hundreds of therapies currently on the market do not conform to these relatively simple standards. This mainly refers to therapies that are carried out in an unconcerned manner by practitioners who show no interest in proving their worth. As for now, their effectiveness should be compared to that of the radio stations installed in "radio homes" in New Guinea. The examples of cargo therapies given below are the most prominent ones which have established themselves well in our tradition, and thus are rarely questioned. Most of them should become the subject of investigations by ethnographers whose field of interest encompasses analysis of cults, myths, rites, superstitions, and other man-made constructs. The fact that what is under scrutiny here is also a construct of people who see themselves as civilized, educated and intelligent, sometimes possessing prestigious academic titles, should not stand in the way of such investigation.

---

[188] C. L. Fincher, and R. Thornhill, "Parasite-Stress Promotes In-Group Assortative Sociality: The Cases of Strong Family Ties and Heightened Religiosity," *Behavioral and Brain Sciences 35*, (2012): 61-79.

[189] Ibid.

[190] There are about 4,200 religions, according to: K. Shouler *The Everything World's Religions Book: Explore the Beliefs, Traditions and Cultures of Ancient and Modern Religions* (Avon, MA: Adams Media, 2010): 1.

[191] T.B. Baker, R.M. McFall, and V. Shoham, "Current Status and Future Prospects of Clinical Psychology: Toward a Scientifically Principled Approach to Mental and Behavioural Health Care," *Psychological Science in the Public Interest 9*, (2008): 67-103.

[192] S. Begley, "Why do Psychologists Reject Science: Begley," *Newsweek*, (January 10, 2009): http://www.newsweek.com/2009/10/01/ignoring-the-evidence.html#

[193] W. O'Donohue and K. E. Ferguson, "Evidence-based Practice in Psychology and Behaviour Analysis," *The Behaviour Analyst Today 7*, (2006): 335-349.

[194] For more about the facilitated communication applied in autistic children see chapter 15 of this book.

[195] http://www.tomaszwitkowski.pl/page7.php?messagePage=3

[196] D. L. Chambless et al., "Update on Empirically Validated Therapies II," *Clinical Psychologist 51*, (1998): 3–16.

Chapter 7: Adult Children of Alcoholics or Victims of the Barnum Effect?

Had it not been for the fact that Phineas Taylor Barnum was born at the beginning of the 19th century – and that he was an honest man – he would surely have risen in the world through presenting himself as a psychological guru. Instead, he was one of pioneers who made a fortune on show business. He certainly did not expect that his name would become enshrined in the canon of psychological knowledge.

Born in Bethel, Connecticut, Barnum started off as a small entrepreneur, with time turning into a public figure and a most charismatic millionaire among the then financial elite, and his circus became the most acclaimed entertainment enterprise in the United States. Famous for the well-known expression "there's a sucker born every minute," which has often been attributed to him,[97] Barnum believed that people deserved to be given more than their money's worth. He was a tireless experimenter that was always on the lookout for things that would bring pleasure to people, and he followed a simple, self-made maxim that show business has "got something for everyone."

His dream was that people would themselves desire to pay for the amusements he placed on offer. To this end, he indefatigably travelled round the world looking high and low for artists and products that would generate rapturous applause from his customers. He used every possible means to attract the audience's attention. Among his many famous characters who provided entertainment to his audience was a dwarf known as General Tom Thumb, the nineteenth century's most outstanding opera singer Jenny Lind, and the amazing Siamese twins Chang and Eng Bunkers. Barnum was always sensitive to people's desires and consequently made sure to provide them with excellent entertainment. He moved heaven and earth to make every member of his public content, and indeed hardly anyone would complain about the quality of performances that he offered.

Before reaching his prime, Barnum had gone through tragedies that might have broken the toughest of the tough. A museum that he had restored to its previous grandeur and cherished as his pride was consumed by fire twice. His mansion, regarded as one of the grandest and most unusual American palaces of the age, also went up in flames. Barnum lost his wife, two children, and got into debt. Despite having suffered such traumas, he never gave up. He pulled himself together in a flash: he grew rich and got to the top one more time, bought new houses, a new museum, and even remarried.

Barnum recounted the story of his life in many biographies that – legend has it – reached the summit of the bestseller lists in the United States, topped only by the Bible. To this day, his life remains a source of incessant inspiration for many businessmen, speakers and artists.[198] His activities stretched far beyond the entertainment market, as he also tried his hand in politics and was a noted speaker, a well-known philanthropist and a skilled socialite. His story would fill the pages of more than one volume. Since an exact account of P.T. Barnum's accomplishments is not the purpose of this study, I will content myself with the necessarily brief summary above, following only one of Barnum's many paths with particularly scrupulous attention. Valid for our discussion here is the aspect of his varied activities that gave him a peculiar nickname, namely the Prince of Humbugs.

As much as Barnum personally did not see anything wrong in using tricks in his circus performances or other forms of amusement provided people were getting value for money, he felt a deep aversion towards those who were lining their pockets by intentionally cheating people. Specifically, he detested organizers of séances and other individuals who promoted themselves as being endowed with mediumistic abilities, the phenomenon that was on everyone's lips at that time. Barnum was even once summoned to court as a witness in the fraud case against a well-known "spirit photographer," a certain William H. Mumler. Without exaggeration it can be stated that P.T. Barnum was a forerunner among those engaged in debunking efforts that would later be undertaken by famous illusionists such as Harry Houdini or the celebrated James Randi, the founding father of the James Randi Educational Foundation. It was none other than Barnum who was the first ever critic to reveal to the public tricks played by spiritualists in order to swindle money out of naive family members who remained in mourning for their deceased relatives. In one of his books (*The Humbugs of the World*, 1865) he offered as much as five hundred dollars, a small fortune at the time, to any medium that was able to prove their ability to communicate with the dead. It was a perfect tool to ridicule alleged supernatural mediumistic abilities. As could be predicted, no one was awarded the prize, just as no one has ever succeeded in getting hold of another reward, this time of one million dollars, that was on offer for several decades under the auspices of Barnum's intellectual successor, James Randi. The long-standing award, known as the Million Dollar Challenge, was to be paid out to anyone who could prove that they possessed paranormal powers in an agreed-upon scientific testing environment.[199] Equally safe was its European equivalent, widely recognized as the Sisyphus Prize. In 2012, one million euros was deposited in the bank account of the Belgian Skeptical Organization (SKEPP) to be raked in by any individual able to demonstrate their paranormal powers under controlled conditions.[200] No claimant has ever made it even through the preliminary testing phase.

But it was not these endeavors that propelled Barnum's name into psychology textbooks worldwide. Before we dig into this, let us try a brief test. Below are two lists of descriptive statements. Read them very carefully and consider to what degree they depict you.

List No. 1
1. You have a great need for other people to like and admire you.
2. You have a tendency to be critical of yourself.
3. You have a great deal of unused capacity which you have not turned to your advantage.
4. While you have some personality weaknesses, you are generally able to compensate for them.
5. Your sexual adjustment has presented problems for you.
6. Disciplined and self-controlled outside, you tend to be worrisome and insecure inside.
7. At times you have serious doubts as to whether you have made the right decision or done the right thing.
8. You prefer a certain amount of change and variety and become dissatisfied when hemmed in by restrictions and limitations.
9. You pride yourself as an independent thinker and do not accept others' statements without satisfactory proof.
10. You have found it unwise to be too frank in revealing yourself to others.
11. At times you are extroverted, affable, sociable, while at other times you are introverted, wary, reserved.
12. Some of your aspirations tend to be pretty unrealistic.
13. Security is one of your major goals in life.

List No. 2
1. You can only guess what normal behavior is.
2. You have difficulty following a project from beginning to end.
3. You lie when it would be just as easy to tell the truth.
4. You judge yourself without mercy.
5. You have difficulty having fun.
6. You take yourself very seriously.
7. You have difficulty with intimate relationships.
8. You overreact to changes over which you have no control.
9. You constantly seek approval and affirmation.
10. You usually feel that you are different from other people.
11. You are either super responsible or super irresponsible – there's no middle ground.
12. You are extremely loyal, even in the face of evidence that the loyalty is undeserved.
13. You are impulsive. You tend to lock yourself into a course of action without giving serious consideration to alternative behaviors or possible consequences.

And now, on a scale from zero to five, where zero indicates very low accuracy and five implies its highest level, rate how well each list describes you. If both description lists seem equally accurate to you, you may ascribe the same rating to them.

Now let us return to the central figure of our story. Phineas Barnum watched closely the work of the fortune tellers employed in his circuses.

What he noticed was people's tendency to accept general characteristics as accurate along with specific descriptions of their own personality, as well as the fact that predictions formulated by psychics would by the same token be frequently interpreted as accurate prognoses of the future. Many years later, his observations were empirically validated by the psychologist Bertram Forer, who even coined the phrase the "Barnum effect," which is used in modern psychology to this very day. This was all the result of an experiment he conducted in 1948 in which a group of students were asked to take a personality test. After the students had filled in the test forms and sufficient time for test evaluation and interpretation had elapsed, he handed out bogus personality profiles to each student and asked them to rate the test results for accuracy. On a scale from 0 to 5, the students on average gave a rating of 4.26. What they had in fact received from Forer was of course an identical collection of hazy statements rewritten from a tabloid horoscope column, one that has been given as the first list above.[201] And now, dear reader, check to what degree your evaluation of the description accuracy tallied with the average score obtained by Forer in his study.

The Barnum effect, also referred to as the Forer effect or the horoscope effect, is observed every time when vague characteristics that define a large number of people are taken as highly individualized and unique diagnoses. As you probably have gathered, this tendency has long been grist to the mill of fortune tellers, clairvoyants, astrologers, spiritualists and other charlatans of all descriptions. While sad to say, it is true that a good number of psychologists can be also found in this group, as will be demonstrated later. Before we identify them, it is worth discussing some other research studies dedicated to the Barnum effect.

An interesting experiment was conducted by C.R. Snyder in 1974. He put participants into three groups. An experimenter claiming to be an astrologer asked participants for their dates of birth. The first group gave no information, and the second revealed only the month and year of their birth. Participants from the third group were asked to provide full dates of birth. All participants then received horoscopes that were allegedly prepared specifically for each person. After the participants had studied those astrological forecasts, they were asked to rate their accuracy on a five-point scale: the more fitting the horoscope seemed to its recipient, the higher accuracy score it should have been assigned. As you may have already realized, all participants were given the same horoscope text. The accuracy ratings proved to be essentially linked to the recipient's conviction that the horoscope had been prepared specifically for them. Those who had not provided the "astrologer" with any personal details before receiving the horoscope gave it an accuracy of 3.24. Participants who had indicated the month and year of their birth considered the horoscope as slightly more accurate, and gave it a rating of 3.76. Needless to say, the accuracy of the horoscopes was rated highest by the third group, that is those who had provided their exact birth date.[202]

Later research additionally demonstrated that, apart from the belief that the description was a person-specific personality analysis, it was also the perceived status of its author that carried the greatest weight when it came to the accuracy assessment. What increased the perceived reliability of descriptions was a high degree of reference to the recipient's positive characteristics.[203]

Therefore, dear reader, if you have considered the characteristics above identified as List No. 1 to be an accurate description of you, you have fallen victim to the very same effect as the majority of participants in the analyzed research studies, and the same one that many years earlier befell the majority of clients who visited fortune-tellers in Barnum's circuses.[204] However, as a matter of fact there is no need to worry.

What may have slightly more severe implications for you is your belief in the high accuracy of the second list. If you have indeed rated those characteristics as highly accurate, this means that you suffer from the syndrome of *adult children of alcoholics* (ACoA)! In point of fact, you should immediately seek contact with the nearest ACoA support group and join them or submit yourself to therapy at your earliest convenience. This is because the many statements that constitute this particular profile for many psychologists and therapists are consistent with the diagnostic profile of individuals affected by this disorder.[205]

Take your time, though, as you may easily end up being entrapped in a modern version of Barnum's circus, only with a different name placed on the signboard at its entrance. This facility will be an establishment of what has become known as the Adult Children of Alcoholics movement. In the United States alone it has attracted nearly 30 million clients.

The roots of this organization lie in *Alateen* groups, which acted as a branch of the Alcoholics Anonymous fellowship (AA). *Alateen* was a community of young people whose lives had been or were defined by the alcohol addiction of a close family member. In 1976 in New York, several people who had previously attended *Alateen* meetings broke away from the group and joined AA meetings as adults. Having failed, however, to find a common language with a community composed of peers of their parents, they decided to set up their own group named Hope for Adult Children of Alcoholics. Whereas the opening and closing procedure for the meeting remained identical as those followed by the AA, the central portion was modified and gave way mainly to personal accounts of the effects of being raised in an alcoholic family on adult life. One day, Tony A., a fifty-year-old member of the AA fellowship, was invited to the meeting as a guest speaker. The age barrier that naturally showed its face at the outset melted away when members of the community opened up to put into words their childhood experiences and emotions. Before long, Tony felt part of the community and became a dedicated member.

Less than six months later, the community suddenly started to shrink and was on the verge of being terminated. In an attempt to save the group from extinction, Tony suggested that it also accept those AA attendees who had alcoholic parents themselves. The first new-format meeting attracted seventeen people, but the third one drew a crowd of more than one hundred and, consequently, the functioning of the group was thus no longer in danger. With the passage of time, a second group was established under the name *Generations*. Tony A. served as the chairperson for both groups. Initially, no formal affiliation requirements were in force, but prompted by other members Tony soon penned a kind of official regulation, known as *The Laundry List*, in which he listed fourteen behavior patterns or characteristics that define the personality of the Adult Children of Alcoholics. This act has been seen as marking the birth of the Adult Children of Alcoholics. *The Laundry List* served as a reference document, based on which similar communities were launched across California and Texas.

No research studies, publications or theoretical concepts have ever been developed or commissioned by the practitioners of the movement to scientifically validate its mission. The driving force behind all its ventures has been the plain, commonsense assumption that a childhood spent in a family with an alcoholic seriously affects adult life. However, as early as in 1979, an article was published in *Newsweek* in which it was first theorized that sharing your life with an alcoholic close family member caused long-term dysfunctional behavior patterns. The first book on this subject, *Adult Children of Alcoholics* by American therapist Janet G. Woititz, was published in 1983 and was an immediate bestseller.[206] Again, in this book empirical evidence is virtually absent. Based on her own observations, the author simply defined a number of characteristics which she thought belonged to the ACoA syndrome. They were subsequently cited and developed by many other writers, likewise away from controlled and structured research. Neither the International Classification of Diseases (ICD) nor the Diagnostic and Statistical Manual of Mental Disorders (DSM) have ever recognized the ACoA syndrome itself as a separate disease or disorder, though it has entered mass culture through the front door and has nested itself in the minds of representatives of Western civilization.

This rapid development of the ACoA self-help movement has coincided (and continues to do so) with independent empirical research. So how does the scientific status of the ACoA syndrome present itself in light of the research? A partial hint is to be found in the work of Mary Longue, Kenneth Sher and Peter Frensch.[207] These psychologists have noted a remarkable similarity between characteristics distinguished as the ACoA personality and descriptions used in studies on the Barnum effect. They decided to test their findings in an experiment in which 224 students at the University of Missouri were invited to participate. They were informed that they were taking part in a scientific project whose aim was to design a new personality inventory.

First, the experimenters divided the participants on the basis of their individual scores obtained in the *Short Michigan Alcoholism Screening Test*. Two groups were formed: one for those identified as ACoA and the other for non-ACoA participants. Then the students answered a series of computer-based questions taken from real personality inventories. Their results were next "analyzed" by a computer, and the scores were supposedly used to generate a personality profile for each examined individual. What every person in fact received was one of four profiles that had been compiled beforehand and then assigned to them on a random basis, two of which were "Barnum profiles", and two ACoA ones. Each of the descriptions had six statements, the accuracy of which the participants were asked to rate.

There were no differences in the evaluations of the profiles between the two experimental ACoA and non-ACoA groups. Both groups confirmed unambiguously that the characteristics described them with a high level of accuracy. Out of those who were randomly given a "Barnum profile," a total of 74% of the respondents (79% of ACoA and 70% of non-ACoA participants) reported that the descriptions had a high or very high degree of accuracy. Sixty-seven percent of all analyzed students (71% of ACoA and 63% of non-ACoA participants) reported that the ACoA profiles described them well or very well. These findings came as a surprise even to the experimenters, who had not anticipated that the accuracy rate indicated for the ACoA profile would be that high and, additionally, that it would be the case irrespective of the actual family background. Moreover, it became apparent that there were practically no significant differences in the perceived description accuracy between the Barnum and ACoA profiles. In the experimenters' view, this evidence precluded the ACoA profile from being a clinical category. They saw ACoA as a harmful myth that stigmatizes people and therefore may bring more harm than good in providing professional support to those in need. Similar findings were presented in 2010 by Kerrie Fineran and associates, which additionally substantiated the claim that the perceived accuracy of ACoA characteristics was simply a natural manifestation of the Barnum effect.[208]

All right, an attentive reader might say, but the only conclusion that may be drawn from the aforementioned research is that tools used for diagnosing the ACoA syndrome are immensely far from perfect. Consequently, one has no grounds to argue that the syndrome does not exist at all. And, indeed, the evidence quoted so far is insufficient to reject the existence of the syndrome, especially given that so many mental health practitioners have committed themselves to help overcome its effects. Let us therefore dig into other studies as well.

An extensive review of research was performed in 1990 by Michael Windle and John Searles in their book *Childrens of Alcoholics: Critical Perspective*.[209] Contrary to claims made by advocates of the ACoA philosophy, the authors established that children of alcoholics did not necessarily become addicts themselves more frequently than study participants from control groups. What is more, as compared to cross-sectional research on the entire population, clinical studies showed a tendency to overestimate the prevalence of familial transmission of alcohol dependence (from parents to children). The review's authors noted a range of other sociodemographic risk factors that might have a greater impact on the emergence of alcohol dependence than just being in an alcoholic family. They also pointed to the role that genetic factors had played in the forming and developing of alcoholism.

No less exhaustive is Kenneth Sher's monograph *Children of Alcoholics: A Critical Appraisal of Theory and Research*, published in 1991.[210] Here again, it was proven that traumas experienced as a result of being raised in an alcoholic family are not determining factors in the person's subsequent, adult life. Sher cautions against the negative impact of labeling people as ACoA.

Other research by Ralph Tarter, Andrea Hegedus and Joan Gavaler demonstrated that sons of alcoholic fathers did not differ from sons of nonalcoholic fathers in being hyperactive. Their findings substantiated the argument that being a member of an alcoholic family did not represent a risk factor in the formation of hyperactivity syndrome.[211]

In 1991, Sandra Tweed and Carol Ryff looked at the psychological adjustment of 114 ACoAs and a group of 125 individuals of comparable sociodemographic background from nonalcoholic families. They found no significant differences between the two groups in most of the analyzed measures with regard to psychological well-being or personality development.[212]

Also in the 1990s, William H. George and associates investigated a sample of 281 individuals for the transmission of alcohol dependence from parents to children. This study was yet another to show no differences between the ACoA group and participants from nonalcoholic families, and many of those identified as ACoA did not display any characteristics of the syndrome.[213]

The body of research is larger and not limited to the studies referenced above. And while the phrase "Barnum Effect" can be found throughout numerous titles and scientific texts, the ACoA movement seems not to lose its appeal. As is claimed by its followers:

> Seventy-six million Americans, about 43% of the U.S. adult population, have been exposed to alcoholism in the family.
> Almost one in five adult Americans (18%) lived with an alcoholic while growing up.
> Roughly one in eight American adult drinkers is alcoholic or experiences problems due to the use of alcohol. The cost to society is estimated at in excess of $166 billion each year.
> There are an estimated 26.8 million COAs in the United States. Preliminary research suggests that over 11 million are under the age of 18.[214]

Could it be, then, that all the research cited above was erroneous and their authors should be regarded as ignoramuses? Such conclusions come to mind after the reading of many texts. Consider just this simple yet deep thought whose meaning I found striking – does it not cast light upon the futility of research?: "Is the ACoA syndrome a fixed set of traits that can be identified in every person raised in an alcoholic family? While it may seem so, it is an utterly untrue statement. *There are as many ACoA syndromes as there are people.*"[215]

Is this not a brilliant idea? Therefore, dear reader, you should now tear out the previous few pages of this book, crumple them up and throw them into a dumpster. Then accuse the author hereof of having led you astray, far off the realm of empirical knowledge. And while doing so, reflect on whether, by any chance, you are an ACoA yourself. If in the United States alone there are 26.8 million people identified as ACoA, it gives us at least 26.8 million syndromes! In finding a therapy that would be suited to your needs (and, I presume, there are also at least 26.8 million therapy models), the Adult Children of Alcoholics® World Service Organization or alternative organizations operating in other countries will come to your aid. The experts whom you will meet there are, unlike primitive empiricists, undoubtedly well trained to detect the syndrome in you and confirm that you have succumbed to its disastrous effects.

A bitter observation? Well, nothing more than another cargo cult! Its priests, just as the Pacific islands inhabitants, seem to be driven by a motto once formulated by Mark Twain: "Having lost sight of our goals, we redouble our efforts." They do more and more research to prop up their concept, and their findings often bring to mind the interpretations made by cargo followers, for whom breezes of a cold wind were manifestations of their deities' presence and treated as a sign of imminent deliveries. Below is a handful of such "chilly breezes."

In a large-scale research study described by Donna Domenico and Michael Windle, ACoA women were analyzed against a control group. It is indeed true that the study found differences to the disadvantage of the ACoA group in almost all study measures; however, most of them were nothing more than purely subjective observations! The ACoA women reported a stronger *sense* of depression and a lower *sense* of self-esteem than the non-ACoA group. It does not stop there. Although the study's authors did not discover any differences in family functioning across both groups, ACoA participants declared a lower *sense* of family cohesion and a lower *sense* of marital satisfaction. Further similar differences in the perceived levels of the analyzed measures were identified. The most striking observation though was that despite the lack of any significant differences in the consumption of substances, ACoA subjects had a *conviction* that they drink more for coping purposes than the control group![216]

Can we treat these findings as indicative of the real psychological disposition of the women under study, or as a hint exposing the bogus reality created and imposed by followers of the "adult child of alcoholic" concept, media in the tank, and a social fad? Already when writing about the myth of childhood, I highlighted the fact that the social sciences shape examined reality to an extent greater than other disciplines of the humanities. Today it is hard to imagine anyone living in an alcoholic family who has not heard about the ACoA syndrome. Sadly, though, the authors of the referenced research did not hit on the idea of investigating the role such an obvious confounding variable as the level of knowledge about the ACoA syndrome might have played. Therefore, we will never be able to establish once and for all whether the differences in the levels of self-perception reported by the study participants sprang by any chance from many years of persuasion by the media or therapists whose aim is to promote their own services.

The findings of research conducted in groups of individuals that have undergone psychotherapy raise yet more doubts. People who are in the middle of the process consistently report subjective symptoms of the ACoA syndrome, with the simultaneous absence of somatic manifestations and no disorders found in their psychological adjustment. Is it still a truthful reality, or an outcome of therapeutic indoctrination? As with the previous questions, there will never be an answer to this one either. But to many ACoA cult disciples, such outcomes are like breezes of a cold wind to cargo cult believers. Even if research-driven conclusions get through to them and reveal that they have been busy lighting up bonfires on nonexistent airstrips, the resultant discomfort will be promptly dealt with.

> Clinical observations made by professional psychotherapists and reliable scientific research demonstrate that many phenomena counted among manifestations of the ACoA syndrome do not affect those diagnosed as AcoA, whereas they do appear in people who have not been raised in an alcoholic family. Nevertheless, I must admit that I myself contributed heavily to the term "ACoA syndrome" being spread in Poland in the late eighties and later. For all that I do not feel remorse about the fact, and I believe that promotion was very helpful at that time. First, it was instrumental in introducing a number of varied forms of support for ACoA, and second in obtaining public funds for the implementation of this aid.[217]

I would instead argue that the promotion of the ACoA syndrome (including by the author of the cited comment) in my country has substantially contributed to state funds being frittered away on treating a disorder that does not exist. This is like someone priding themselves on having acquired public money to finance support given to those who had been given a bad horoscope from an astrologer. Is this not similar to the explanation that Stanislav Andreski, cited earlier in the book, had in mind when depicting the line of reasoning followed by a representative of the social sciences who suddenly realizes the futility of his own exploits?

"His dilemma, however, stems from the difficulty of retracing his steps; because very soon he passes the point of no return after which it becomes too painful to admit that he has wasted years pursuing chimeras, let alone to confess that he has been talking advantage of the public's gullibility. So, to allay his gnawing doubts, anxieties and guilt, he is compelled to take the line of least resistance by spinning more and more intricate webs of fiction and falsehood, while paying ever more ardent lip-service to the ideals of objectivity and the pursuit of truth."[218]

The process of fueling the faith in cargo pursues multiple paths. The aforementioned interpretations of research findings are not the sole manner of strengthening the cult. Similar to the prophet Gogol mentioned above, the ACoA priests are also highly skilled in transforming their earlier prophecies when faced with an inescapable defeat. For instance, faced with a growing number of studies proving that mental health issues experienced by people who have been "diagnosed" as ACoA do not in principle differ from those suffered by most of us, they swiftly put different data on the table, with a view to extending the field of their impact and reinforcing the cult itself.

In 1997, Christine Harrington and April Metzler conducted research whose aim was to answer the question of whether ACoA was something distinct from *adult children of dysfunctional families* (ACoDF), that is adult children of dysfunctional families with no alcohol dependence.[219] Not only did this analysis find no differences between those two groups, but additionally the ACoAs did not differ from the control group composed of participants from functional families. The only statistically significant differences found were identified between both experimental groups, ACoA and ACoDF, and the control group with regard to communication difficulties in problem-solving tasks. And nothing more than that!

What conclusions may be drawn from this? Well, there is a very obvious one for a person of average intelligence who knows the fundamentals of logical inference. If ACoA and ACoDF do not show any dissimilarities and they additionally do not differ much from the rest of the population, there is not much point in establishing such syndromes as separate mental health categories. What do such results mean to cargo cult priests? Nothing else than further extension of their sphere of influence. Not only do they not intend to give up their existing followers, now they may extend their reach even farther and deliver their prophecies to those individuals who were brought up in families that were not alcoholic! A dysfunctional family is obviously one which is ruled by domestic violence, but also one where the child was raised by one parent or where parents were too busy with their work and ignored their children, or even where parents had an overly pedantic approach to domestic routines.

If a target group that lies within the interest of ACoA therapists and activists accounted for 43% of the adult population of the United States until today, just imagine what a large audience has just opened up to them! Fifty percent of the population? Sixty? Perhaps more. Is it not a heaven-sent message for the cargo priests? From now on they are free to speak unreservedly about ACoA and ACoDF problems.

> The "Adult Children Of Alcoholics" title can be a bit of a misnomer. It is a blanket term that includes those who suffered abuse or neglect even though their parents didn't drink. The cause of the trauma matters not. Everyone who suffered from dysfunctional parenting is welcomed into the fellowship of ACA with open arms.
> My fear is that some of these people may never make it to an ACA meeting. They may, after hearing the word "alcoholic," quickly decide that they do not belong in this group. I have the advantage of knowing that even if my parents stopped drinking it would not have changed their behavior. The anger, judgement and fear I grew up with would still have emerged.
> Recently I saw the term "Adult Children of Dysfunctional Families," or ACODF. It seems a more inclusive term than ACA or ACOA and more clearly describes the broad spectrum of people who grew up with all types of parental dysfunction. This acronym is a more apt fit for the majority of the population that grew up with dysfunction, with or without the substance abuse component.
> There are many who don't know where to seek help or have yet to discover they even need it. Here's hoping the inclusive term, ACODF, will inspire a feeling of belonging and is seen as a warm invitation to join in recovery![220]

And yet this is not the whole story. The ACoA cult priests consistently enlarge their target group. In doing so, they make greater use of the evangelizing practices of the West than the religious experiences of the New Guineans, as can be presumed from another of their discoveries, that of codependency. Therapists had already coined this term in the 1970s, but it came into popular use only later due to Janet Woititz, who has been spoken of earlier in the book. With time it was also accommodated by the AA movement. The acclaim of books describing "toxic relationships" caused the term to be uncritically applied to label anyone in whose family there was or is an alcoholic.

When Woititz's books made a splash in the market and consequently turned out to be true bestsellers, an animated debate sprang up over psychological processes that run in alcoholic families. Addiction professionals at that time shared a fairly prevalent conviction about the momentous character of those discoveries that, following the author, were most often spoken of in terms of new disease entities. It seemed only a matter of time before the new nosological entities were admitted into the domain of psychiatric classification. Yet, not only did these hopes go unfulfilled, they also grew considerably more distant.

The absence of a disease category is seen as no obstacle in conducting therapy by psychotherapists. Therefore, codependency has become a subject of interest for countless institutions and therapy centers. As is clear, the profession has succeeded in first appropriating ACoDF as a natural offshoot of ACoA, and then in seizing all codependents.

> Never before in the history of Twelve Step programs has a fellowship brought together such a diverse group of recovering people that includes adult children of alcoholics, codependents, and addicts of various sorts. The program is Adult Children of Alcoholics. The term "adult child" is used to describe adults who grew up in alcoholic or dysfunctional homes and who exhibit identifiable traits that reveal past abuse or neglect. The group includes adults raised in homes without the presence of alcohol or drugs. These ACA members have the trademark presence of abuse, shame, and abandonment found in alcoholic homes.[221]

How far is this from having the entire population referred to therapy? We are perhaps not there yet, as cargo cult priests are still engrossed in a struggle to magnify the area of their influence. To this purpose, they reach for another syndrome, that is Fetal Alcohol Syndrome (FAS), or in fact for its outgrowths. The term was inserted into the clinical jargon in 1973 by two American doctors, Kenneth L. Jones and David W. Smith. Today, it has turned into a separate diagnostic entry to be found in the ICD-10 recognized diagnostic system that incorporates physical and mental disorders believed to spring from the effects of alcohol on a child in the prenatal period. It is estimated that around 1% of newborns suffers from FAS.[222]

The very existence of this syndrome cannot be questioned and I am aware of no one who would do so. Take heed, though, as the creativity of therapists knows no limits. New clinical syndromes are popping up like mushrooms and, as we might readily imagine, they happen to be a perfect fit for therapy:

- FAE (*fetal alcohol effect*) – a neurological impairment that has the same symptoms as FAS but is of lesser severity. With no distinctive morphological features present, it is recognized exclusively on the basis of diagnosis of non-somatic disorders and is subject to verification through medical history;
- PFAS (*partial fetal alcohol syndrome*) – partial FAS;
- ARND (*alcohol-related neurodevelopmental disorder*) – a disorder that encompasses a range of symptoms solely in mental (learning difficulties, poor memory) and social abilities (stunted social relationships); physical abnormalities are usually absent;
- ARBD (*alcohol-related birth defects*) – a common term for the full spectrum of fetal disorders that result from exposure to alcohol during fetal life;
- FASD (*fetal alcohol spectrum disorders*) – describes the entire range of disorders related to the exposure to alcohol in the womb.

Except for FAS, none of the aforementioned syndromes is recognized as a separate diagnostic entity. There are hardly any differences between some of them, as in the case of ACoA and ACoDF. Their originators should direct their attention to the Occam's razor principle: *entia non sunt multiplicanda praeter necessitatem* (do not multiply metaphysical entities without necessity). But, to be clear, this is not about the economy or simplicity of reasoning. The thing at stake is the sphere of influence.

Diagnosing FASD, FAE, PFAS, ARND or ARBD is a process founded on much vaguer and typically behavioral measures, as opposed to the well-defined symptoms of FAS.

Alcohol addiction and its repercussions can be a real nightmare for dependent persons, their future children and other family members. Many of those addicted are without doubt in deep need of professional psychological assistance. Alcohol dependence represents one of the most complex problems of our civilization. What is most disgraceful, however, about the actions taken by cargo priests is that these specialists do not aim their efforts at solving the problem of alcoholism. The bulk of the propaganda campaigns, publishing activities and therapeutic measures that I have written about do not focus on the very problem itself, but only on its alleged consequences. Let me emphasize once more – alleged consequences. The more people are enticed to believe that their life's problems have arisen from the syndrome of ACoA, ACoDF, FASD, FAE, PFAS, ARND, ARBD or codependency, the stronger the cult grows and, following that, the higher the social status of its priests becomes.

It is with full awareness of this that I have written that many victims of alcoholism need psychological support. Likewise, my use of the term "professional" was intentional. Not just any support, or common-sense, ideological or pseudoscientific. The ACoA fellowship and other referenced movements may be accused not only of being engaged with something that escapes diagnosis and of blowing up problems to an unimaginable degree. If it were only this issue, one could just dismiss it, assuming that perhaps, by dealing with millions, they have managed to help at least a few. Unfortunately, while support available under the aegis of the ACoA movement is anything but professional intervention, it is also heavily loaded with indoctrination based on erroneous underpinnings.

The inhabitants of New Guinea, taken in by promises made by the cargo cult priests, abandoned their daily routines, crop cultivation, animal breeding or, from time to time, threw themselves into a frenzy of spending money. The ACoA cult follows a similar path. Myriad people spend – and not infrequently waste – their lifetime on explorations that will not necessarily lead them to solving their problems. Many must feed on false self-images or are sentenced to working out non-existent problems.

The approach acclaimed by the ACoA movement and affiliated therapists is a peculiar brew of common sense, half-truths and common observations on psychology, metaphors, false assertions and harmful recommendations. Take, for instance, just the first mantra of the whole movement, that is, the well-known conviction about the necessity of delving into and retrieving childhood events as a prerequisite for sorting out one's adult life. This is how Janet Woititz, the author of the ACoA "bible," openly addresses this issue: "I hope that you are reading this book with the intent of airing some issues, so that wounds will be lanced and disinfected, knowing that they will then have the opportunity to heal."[223]

This message reverberates in the words of another DDA prophet, Timmen Cermak:

> It is very helpful to look carefully and honestly at the past if you want to make changes in your life. Remembering the past – both the factual details and the emotions that colored each moment – usually gives us a more sympathetic perspective on our problems. ...
> Once memories have been buried alive, without a headstone, we can lose track of where they lie.
> As long as you remain in the straitjacket of not talking to others about your experience and disowning your feelings about the past, memories may remain few and far between.[224]

How highly ineffective and even destructive such ideas may be was examined and highlighted in my previous book, *Psychology Gone Wrong*, where we unfolded the secrets of therapies based on childhood analysis.
Understanding human emotions, which is seen as one of the mainstays of the ACoA movement, also has little to do with modern psychological knowledge. This is more of a mixture of psychoanalytical views on the hydraulic model of emotions, as well as selective clinical observations. Below is a sample of this joyful creativity:

> Sometimes the unresolved anger leaps out with unexpected intensity at a spouse or one's own children. Sometimes ACoAs create emotional distance from others when they refuse to acknowledge their anger.
> Intimacy is blocked by unresolved anger because anger cannot be banished from our lives without becoming a dark shadow that relentlessly follows us. In Jungian theory the "shadow" is made up of all those facets of our personality that we find too uncomfortable to own up to. The shadow that we all possess also possesses us. When people deny anger its rightful place, they experience the dark side of their disowned anger in two ways: First, they have no choice but to be "nice" to others. Their own feelings count for nothing. But they find that the feelings remain alive no matter how long they have been buried, and they continually find cracks and back alleys to squeeze through.[225]

Spinning stories about "blocked intimacy," "shadows," "cracks," "back alleys" and other similar nonsense causes many people to misinterpret emotions and make needless and vain attempts to alleviate them. Such viewpoints are alarmingly close to the glorification of anger – an emotion that has gained odd recognition in ACoA.

> It is good to feel anger because it gives us strength and determination to persevere in the struggle for what is important to us, to restore order where it was once shattered. It is a source of enormous power. Under its influence, a person is able to defeat an assailant tougher than themselves, to endure an even worse situation and yet win, drawing on the faith that they are the ones who are right.
> The fiercer the anger, the greater the likelihood that the past wound was indeed severe. Yet, it triggers an impulse to disinfect the wounds, even if it translates into going through a very painful process. And this means that anger may be a constructive feeling to be used as a weapon of self-defense, rather than as a weapon of attack.
> Constructive use of anger is one of the social skills. It gives opportunities to make a lasting change in ourselves, our surroundings and memories. It awakens a surge of might and provokes a sense of empowerment.[226]

I am aware of no empirical study that would support a claim that the volume of anger is dependent on the severity of the trauma experienced. Likewise, I have not seen any scientific research findings that would demonstrate the surge of "might and a sense of empowerment" under the influence of anger, though I am sure that there are a fair number of anecdotes on the subject. To the contrary, many research reports can be found to prove that unloading negative emotions leads to the escalation of aggressive behavior.[227]

Still, this is not the end of all the negative outcomes that participation in the ACoA scheme may bring about. Jeffrey Burk and Kenneth Sher have done extensive research into the adverse effects of labeling people as ACoA. Above all, participants involved in ACoA self-help groups proved to be prone to developing hectic and complicated interactions with parents and family role relationships. Social stigmatization of such individuals was also not rare. Many of them shaped self-images that were thoroughly distorted, yet compliant with the ACoA concept, and they misinterpreted the underlying causes of their physical and mental issues.[228] Experimental studies also corroborated the strong negative effects of being labeled as ACoA.[229]

The most profound lure of the ACoA movement, being at the same time its greatest hazard, is the belief that we are not ourselves responsible for the quality of our adult life. This is what people are taught there: "Many people are unaware that they should not always be blamed for failures that come their way. Our breakdowns or the lack of success can frequently be attributed to the experience picked up in a dysfunctional home."[230]

Imagine you have formed and strengthened such a belief, and, subsequently transferred it onto your whole life and life circumstances. This is a most expedient path to not discovering the true causes of your life mistakes. If we make people believe such theories, we teach them that there is nothing they can do to improve the quality of their life.

Being mindful of the force with which the ACoA movement is being cultivated, as well as its emotional ideology, I know for sure that the charges I have levelled above will come under fierce criticism. Much of this I have already experienced, as I heard it repeatedly during my lectures and meetings with my readers. Well, this line of polemics was already explored by Arthur Schopenhauer in his book *The Art of Being Right* (*Die eristische Dialektik*). Since it is always worthwhile to allude to the classics of philosophy in your argumentation, let us analyze the potential attacks while already being armored with the skills of polemical sparring. This will enable us to save time and avoid a futile dispute.

Without doubt, one of the first objections will be cleverly hidden in the question: "Are you ACoA yourself"? No, I am not. And to the next "argument," namely "So how can you play a know-it-all and comment on sufferings of those who have been affected by such a traumatic past?", I shall reply: "One does not need to be a bird to possess knowledge about flying. An average ornithologist has more to say about flying than an average bird. A scientist does not have to turn into a crystal so as to be able to look into its structure. The ACoA syndrome is exactly the same area of study as any other."

Next, I will most likely be inundated with evidence given by "healed and happy ACoAs". To this dictum I shall explain that many "healed," "happy," "saved" and other people miraculously rescued from life's adversities are to be found in each religious sect worldwide, not excluding the New Guinea cargo cults. In the same way, one could point out examples of many "healed" patients who went through homeopathic treatment and other charlatanic procedures, many who are highly satisfied with their horoscopes received from astrologers, and also those for whom "prophecies" made by clairvoyants have come true. Such evidence can be treated only as proof of how common and powerful the defense mechanism of rationalization is, but on no account does it have value as scientific data. As a matter of fact, it can also be easily denied by a copious amount of evidence to the contrary, such as that given by one of my blog's readers: "Can you imagine that when I learned that I was ACoA and ACoDF, I wasted nearly five years explaining and digging up dirt from the past?"

To all of you who can "see" benefits that participation in ACoA brings, whether as patients, attendees of self-help groups or through a therapist's eye, let me remind you of an old though not always evident truth: the earth is round, not flat. And although for many years my senses have been experiencing its "flatness" every day, I know thanks to knowledge based on scientific evidence that the reality is different. And I am able to accept this truth.

While watching the frantic development of ACoA and related syndromes, I have invented another, equally lucrative business, that is Adult Children of Psychologists. Now, if your parents are psychologists, who use stereotypes on the functioning of psyche on a daily basis, it is simply impossible for you grow to up and be normal; this is my claim. Thus, there surely must exist the syndrome of Adult Children of Psychologists (ACoP), which can only be overcome with assistance from a support group. Similar to the ACoA concept, mine has correspondingly solid empirical grounds. But this should not disturb us. You are welcome to join us! We will set up some code together and establish a procedure of several-something steps. We will cheer one another on, encourage others to sign up and publish articles that explore the subject. The best among us will be chosen as chairpersons of our group. We will create our own, unique jargon and will cut ourselves off from critics who are unable to understand our particular position. My dear ACoPs, to attain maturity and autonomy, you should... But hang on! I will only reveal what you should do after you have signed up for my group

---

[197] According to http://skepdic.com/barnum.html, Barnum did not originate the expression. David Hannum, the leader of a syndicate that had purchased the Cardiff giant, was quoted as saying "There's a sucker born every minute" when he heard of Barnum's plan to display a fake of the fake giant.

[198] See: J. Vitale, *There's a Customer Born Every Minute: P.T. Barnum's Amazing 10 "Rings of Power" for Creating Fame, Fortune, and a Business Empire Today -- Guaranteed!* (Hoboken, NJ: Wiley, 2006).

[199] More about *One Million Dollar Paranormal Challenge*: http://www.randi.org/site/index.php/1m-challenge.html.

[200] http://en.wikipedia.org/wiki/SKEPP#The_One_Million_Euro_Sisyphus_Prize

[201] B. R. Forer, "The Fallacy of Personal Validation: A Classroom Demonstration of Gullibility," *Journal of Abnormal and Social Psychology 44*, (1949): 118-123.

[202] C. R. Snyder, "Why Horoscopes are True: The Effects of Specificity on Acceptance of Astrological Interpretations," *Journal of Clinical Psychology 30*, (1974): 577-580.

[203] C. R. Snyder, R. J. Shenkel, and C. R. Lowery, "Acceptance of Personality Interpretations: The "Barnum Effect" and beyond," *Journal of Consulting and Clinical Psychology 45*, (1977): 104-114.

[204] The first list of characteristics was taken from: Forrer, "The Fallacy."

[205] The second list was created on the base of 13 traits of ACoA published by J. Woititz, *The Struggle for Intimacy* (Deerfield Beach, FL: Health Communications, 1985), 109-130.

[206] J.G. Woititz, *Adult Children of Alcoholics* (Deerfield Beach, FL: Health Communications, 1983).

[207] M. B. Logue, K. J. Sher, and P. A. Frensch, "Purported Characteristics of Adult Children of Alcoholics: A Possible 'Barnum Effect,'" *Professional Psychology: Research and Practice 23*, (1992): 226-232.

[208] K. Fineran, J. M. Lauxb, J. Seymourb, and T. Thomas, "The Barnum Effect and Chaos Theory: Exploring College Student ACOA Traits," *Journal of College Student Psychotherapy 24*, (2010): 17-31.

[209] M. Windle, and J. Searles, *Childrens of Alcoholics: Critical Perspective* (New York: Guilford Press, 1990).

[210] K. Sher, *Children of Alcoholics: A Critical Appraisal of Theory and Research* (Chicago: The University of Chicago Press, 1990).

[211] R. Tarter, A. Hegedus, and J. Gavaler, "Hyperactivity in Sons of Alcoholics," *Journal of Studies on Alcohol 46*, (1985): 259–261.
[212] S. H. Tweed, and C. D. Ryff, Adult Children of Alcoholics: Profiles of Wellness Amidst Distress," *Journal of Studies on Alcohol 52*, (1991): 133–141.
[213] W.H. George, J. La Marr, K. Barrett, and T. McKinnon, "Alcoholic Parentage, Self-Labeling, and Endorsement of ACOA-Codependent Traits," *Psychology of Addictive Behaviors 13*, (1999): 39-48.
[214] National Association for Children of Alcoholics, http://www.nacoa.net/impfacts.htm
[215] J. Mellibruda, "Trzy miliony DDA," *Charaktery*, 2, (2011): 74-77, emphasis mine.
[216] D. Domenico, And M. Windle, "Intrapersonal and Interpersonal Functioning Among Middle-Aged Female Adult Children of Alcoholics," *Journal of Consulting and Clinical Psychology 61*, (1993): 659-666.
[217] Mellibruda, "Trzy miliony DDA."
[218] Andreski, *Social Sciences*, 24.
[219] C.M. Harrington, and A.E. Metzler, "Are Adult Children of Dysfunctional Families with Alcoholism Different from Adult Children of Dysfunctional Families Without Alcoholism? A Look at Committed, Intimate Relationships," *Journal of Counseling Psychology 44*, (1997): 102-107.
[220] "ACODF," *Adult Children Of Alcoholics/ ACAs ACOAs ACODFs Blog*, (September 20, 2011): http://adultchildrenaca.blogspot.com/2011/09/acodf.html
[221] "Welcome To Adult Children Of Alcoholics®/Dysfunctional Families," *Adult Children Of Alcoholics® World Service Organization*, http://www.adultchildren.org/
[222] P. A. May, and J. P.Gossage, "Estimating the Prevalence of Fetal Alcohol Syndrome. A Summary," *Alcohol Research and Health 25*, (2001): 159–67.
[223] J. G. Woititz, *The Complete ACOA Sourcebook: Adult Children of Alcoholics at Home, at Work and in Love* (Deerfild Beach, FL: Health Communications, Inc., 2010).
[224] T. L. Cermak, and J. Rutzky, *A Time to Heal. Workbook* (New York: G.P. Putnam's Sons, 1994), 48, 52.
[225] Ibid., 132-133.
[226] M. Kucińska, „Złość ma swoje dobre strony," *Charaktery*, 12, (2006): 72-73.
[227] B. J. Bushman, R. F. Baumeister, and C. M. Phillips, "Do People Aggress to Improve Their Mood? Catharsis, Relief, Affect Regulation opportunity, and aggressive responding," *Journal of Personality and Social Psychology 81*, (2001): 17–32; W. A. Lewis, and A. M. Bucher, "Anger, catharsis, the reformulated frustration-aggression hypothesis, and health consequences," *Psychotherapy 29*, (1992): 385–392; J. Littrell, "Is the Re-Experience of Painful Emotion Therapeutic?" *Clinical Psychology Review 18*, (1998): 71–102; R. H. Hornberger, "The Differential Reduction of Aggressive Responses as a Function of Interpolated Activities," *American Psychologist 14*, (1959): 354.
[228] J. P. Burk, and K. J. Sher, "The "Forgotten Children" Revisited: Neglected Areas of COA Research," *Clinical Psychology Review 8*, (1988): 285–302.
[229] J. P. Burk, and K. J. Sher, "Labeling the Child of an Alcoholic: Negative Stereo-Typing by Mental Health Professionals and Peers," *Journal of Studies on Alcohol 51*, (1990): 156-163.
[230] http://www.opowiadania.pl/sprint.php?item=4890

Chapter 8: Between Life and Death: The Simonton Method

At least one in four, and closer to one in three readers of this book, will die of cancer. This is a simple statistic that leads rationally thinking people to treat such a possibility as very likely. And this is what many do, trying to live a lifestyle minimizing this risk to some degree. Unfortunately, we are powerless to influence the overwhelming majority of cancer-inducing factors, and there are still others which simply remain unknown. With the present state of developments in medicine, many of us receive our death sentences far before they are actually carried out. A diagnosis of some types of cancer, AIDS, or other such untreatable diseases frequently constitutes such a sentence.[231] Support from our surrounding environment is exceptionally important in such moments. We treat doctors as oracles, and we are inclined to perceive superhuman capacities in them; even a nurse, at an opportune moment, can transform into an angel of hope. It simply doesn't cross our minds that someone could be indifferent to what is happening to us; we can't imagine that we might be treated as objects; and we are entirely incapable of thinking that someone might cynically take advantage of our situation.

In these circumstances, terribly trying both for the sick person as well as members of the family, psychologists are the bearers of hope – people who can sooth the souls of the afflicted and fuel the heart with optimism. Many of them have dedicated their entire lives to helping people with cancer. They have developed a new field of knowledge that examines such subjects as the links between the course of cancer and psychological factors. The founder of psycho-oncology, which is the name given to this new science, is Jimmie Holland, chair in Psychiatric Oncology at the Memorial Sloan-Kettering Cancer Center.[232] She has defined psycho-oncology as an interdisciplinary sub-specialization of oncology that researches the emotional reactions of patients in various stadia of cancer, along with their families and the medical personnel engaged with them. Psycho-oncological knowledge is used today as a means of helping patients in a manner appropriate to a given phase of treatment (and illness). The primary forms of aid offered are psychoeducation, support, attitude changing, and debunking myths associated with cancer. This is done with the use of methods and techniques applied in psychotherapy.

However, in spite of how psycho-oncology is the domain of many outstanding scientists, it has been dominated by one figure – Carl Simonton. At least, this is true of the version of psycho-oncology popularized in the media and the Internet. I know that a Google search is not a precise method of measuring anything, but it can most certainly help us in quickly grasping the popularity of certain names and events. In the case of Carl Simonton, Google generates around 59,200 results, compared to just 19,700 for Jimmie Holland. Both names together occur on just 808 pages.[233]

This is interesting, as we may obtain an entirely different picture of the discipline when searching scientific databases. In two of them, PsychINFO and MEDLINE, Jimmie Holland is listed as the author or co-author of thirty-two scientific publications, twenty-nine of which have been published in refereed journals, meaning those of significance in the scientific community. The name Carl Simonton, however, only appears three times, yet on the cover of the Polish edition of his book he is presented as the "founder of psycho-oncology."[234] Where did Carl Simonton come from, and how has he achieved such incredible popularity around the world while remaining almost unknown in the scientific community?

The web page of the Simonton Cancer Center informs us that our hero is a graduate of the Oregon Medical School, after which he worked in oncology as a radiotherapist for three years. Later, as the head of Radiation Therapy at Travis Air Force Base, he began to implement his own model for working with patients.[235] As we can read, he achieved incredible results immediately:

> The first patient was a 61-year-old male with an advanced throat cancer. He lost over 30 pounds, he could barely swallow his own saliva and had difficulty breathing. He was expected to get worse despite treatment. Carl taught this patient to imagine a desirable outcome, imagining his cancer as curable and his treatment as his allies and friends and as being effective and that his body was capable of overcoming cancer. What was the most astonishing was that the patient had no side effects to high dose radiation therapy. This was also an example that despite the patient's criminal past he could get well. Carl emphasized, "You don't have to be a saint for a miracle to happen to you."[236]

Of course, it is thanks to Simonton's help that the patient had a full recovery, but unfortunately nobody in the medical literature apart from Simonton wrote up a description of the case, and the beginnings of the brilliant therapy are lost in the fog of the past. So is it true or false? I am afraid that Simonton took his secrets with him to the grave. Interestingly, the majority of miraculous cancer healers provide similarly shocking examples, and the majority of them have experienced serious difficulties in providing us with details.[237]

Where did the Simonton method come from? Did he, as most geniuses do, develop it himself? At any rate, the Simonton Center's web page does not inform us that during the 1960s and '70s he attended a course in *Mind Dynamics* led by Alexander Everett, the British personal development consultant and author of motivational books, who himself developed his methodology primarily on the basis of techniques created and propagated by one José Silva.[238] I have dug so deeply into the sources in order for you, dear reader, to have a full picture of the method under discussion, because the truth differs a bit from the marketing.

José Silva was born in Laredo, Texas, in 1914. During his childhood, circumstances forced him to take responsibility for feeding his family, and so he engaged in various kinds of work. When he began working as a radio mechanic he finally succeeded in achieving some financial stability, which then allowed him to engage in parapsychology and occult experimentation. While he was developing the foundations of his method, he made contact with the pioneer of parapsychology experiments, Joseph Banks Rhine, but he rejected Silva's work. Even for a supporter of parapsychology, albeit one with a certain sense of methodological discipline, Silva's explanations seemed outrageous; indeed, he claimed that his teacher had been "a Chinese spirit guide sitting in Yoga position in the astral plane, whom he met at the beginning of his career, practicing the art of out-of-body experience."[239]

This information is invaluable, as from the moment when Silva was given this priceless gift from his Chinese spiritual guide, the method has undergone an exceptional transformation. Presently, we may read on web pages promoting the method that: "The Silva Method is a course in self-control of one's mind during which attendees learn how to employ modified mental states in order to effectively solve problems. What distinguishes the Silva Method from other techniques and methods of mental self-control is that from its very beginning (in 1944) it has been based on a firm scientific foundation."[240]

Reading such things drives me batty – I have never come across a methodology textbook that would make mention of a Chinese spiritual leader sitting in a yoga position on the astral plane. I have no idea what José Silva was drinking or smoking before he met with this said Chinese person, but in spite of the colorful descriptiveness of his vision he is incapable of covering up certain deficiencies in his education – as far as I know, the Chinese do not have much in common with yoga, apart from being "from the East." As we can see, this is far enough from Texas that certain things could imperceptibly blend together and seem deep and piercing. (Just to be sure, I searched the articles in the EBSCO database. The name José de Silva does appear in several places as the author of a publication, but never is it the José Silva we are speaking of here.)

So we have already familiarized ourselves with the theoretical pre-history of Carl Simonton's method. He himself does not make any secret of it. During the Silva Method International Convention in 1974, he gave a lecture during which the Silva Method came in for deep praise. Among other statements about the method, he said "I would say it is the most powerful single tool that I have to offer patients."[241]

The texts we may read on web pages propagating the Silva and Simonton methods are striking, however, owing to a certain asymmetry. Nearly every site devoted to the Silva Method contains information about Simonton's devotion to the master, and he is named practically everywhere as a pupil-continuer of Silva. However, it is impossible on pages devoted to Simonton to find references going in the other direction. Simonton's biograph contains no information about training in *Mind Dynamics* or about his participation in conventions focused on theory and organized by Silva adherents. Why? The answer is rather simple. Both methods aspire to wear the label of "science," but while Silva, a simple electro-mechanic, failed to cross the threshold of science, the method developed by Simonton, who, in the end did possess a medical education, has made fantastic strides into an area that at least has the trappings of science. Social psychologists, specialists in exerting pressure on others, speak in these cases of strategies for "basking in the glory of others" and "escaping the shadow of failure." For supporters of Silva, the invocation of Simonton is a way of bathing in his light, while the status of Silva's method in the scientific community leads supporters of Simonton to run away from the shadow of failure.

The sources of inspiration that give rise to a given method are very important, but they do not, in and of themselves, determine the worth of the direction in which it has developed for almost 40 years. Perhaps Dr. Simonton succeeded in rejecting initial, mistaken assumptions and developed an effective method? It is worth having a look, particularly considering that supporters of this concept advertise its exceptional effectiveness.

> The Simonton Program, which is presently taught in many European countries (including Poland), the USA, and Japan, was the first psychotherapy program proven in scientific studies to successfully complement conventional treatments with appropriate psychotherapeutic intervention, and on average doubles the survival rates of the ill while providing them with significant improvements in quality of life.[242]

> The Program's Effectiveness: patients who participated in psychotherapy along with traditional treatments lived approximately twice as long, more of them made a complete recovery, they experienced better quality of life, and if they died, they also experienced better quality of death.[243]

When journalists write about these revelations, this is how it sounds: "In his work he follows the method developed by the American oncologist Carl Simonton, who was the first doctor to incorporate psychotherapy into treatment for cancer patients. His recommendations provoke skepticism among some, but studies have demonstrated that patients treated using the Simonton method saw their cancer cured twice as frequently as those treated exclusively by traditional medicine. And among the ill who died, Simonton's patients lived twice as long."[244]

I asked the author of the interview cited above for the source of her revelations. In response, I was told that she had worked on the text a long time ago, and she sent me the phone number of someone who was supposed to know something about those studies. That person, in turn, referred me to two sources – websites describing the Simonton program and presenting writings by the very founder of the method, and to Ewa Wojtyna, a psychologist supposed to have conducted research on the subject. Indeed, she does engage in research, but on coping with chronic back pain. A doctor acquaintance asked for the same information from Inka Weksler, the founder of Lower Silesia's first psycho-oncology clinic. She also was unfamiliar with the sources, but in the conversation cited above she did not correct the information given by the journalist. If you, dear reader, are now wondering whether it is possible to conduct therapy without even the foggiest notion as to the results of studies on the effectiveness of the method employed, you now have an answer – it is possible! This is even a quite widespread phenomenon, at least in Poland.

Ms. Weksler was kind enough to provide us with the e-mail address to the medical director for the PsychoSocial Oncology program in the Todd Cancer Institute at Long Beach Memorial, Mariusz Wirga – presently one of the most ardent supporters of the Simonton method. As it was, he turned out to be a bit more informed, and he told us the following:

> The above statement is not precisely formulated. First and foremost, all of the patients in those studies were in an advanced stage of metastasizing cancer. Simonton's patients compared to those treated in the best cancer centers in the USA survived on average twice as long, and among them were far more long-term survivors (no sign of the disease after 5 years). They experienced better quality of life, and those who died had a better quality of death. The results of these studies are described in Simonton's book *Getting Well Again*, as well as in the articles cited below.[245]

I read over both articles, and at the beginning I analyzed the one whose abstract referred to the doubled life expectancy of Simonton's patients.[246] Because I had never conducted any epidemiological studies, I asked a friend and cancer specialist for help. The presented analysis thus contains elements that are of importance in general research methodology and statistics, as well as more detailed medical issues.

The study's authors made use of a sample of 225 patients suffering from breast, intestinal, and lung cancers. The patients were given both conventional treatment and psychotherapy according to the Simonton method, and life span was compared to the standard survival rates for patients experiencing breast, intestinal, and lung cancer in leading cancer treatment centers. The authors drew their conclusions as to the effectiveness of the Simonton psychotherapy method based on the differences between the observed life expectancy of patients and data from the literature.

Firstly, this type of study is not an experiment, as its authors contend. The procedure described in the preceding paragraph may, with a healthy dose of goodwill, be described as a quasi-experimental study plan. What is missing is a control group, a randomized assignment of people to groups, and real controlling for independent, dependent, and other variables. In research methodology, particularly when it is difficult to find research groups (or random assignment to groups would be unethical), quasi-experimental plans are frequently accepted, and even recommended as the only possible ones in such a situation. However, the study by Simonton and his collaborators contains a large number of errors which could be eliminated without any detriment to the patients. First and foremost, we have no information concerning patients from a comparison group. We do not know at what stage of the cancer they received their diagnosis, we do not know when treatment was initiated, we do not know their ages, and there are many other pieces of information we do not know that have significant impact on survival rates. We also do not know the basis on which Simonton's patients were qualified for therapy and for the study. Was every person afflicted by cancer in the given area provided with the therapy? Were they handpicked patients who were in a private clinic? Most importantly, we do not know whether the medical procedures applied to the patients from the experimental group and the comparison group were the same. After all, medical progress has the strongest influence on the results of treatment. As I have already mentioned, all of these flaws could have been remedied without detriment to the lives of the patients.

This is not all – in their calculations concerning the life expectancies of Simonton's patients, there is no account taken of people who died in the course of therapy or prior to its initiation, while such data is generally taken into consideration, at least in national statistics. Also surprising are the ways in which some data are calculated. For example, the authors compare some of the measures of central tendency with each other – the median. This is a value taken from the analyzed data set above and below which exactly half of results are located. Let us assume that we have patients who survived 1, 1, 2, 3, 9, 10, 10, 10 and 10 months, respectively. The median of this set is 9. It is easy to observe that this is an entirely different value from the mathematical average, which is 7.22 (and thus the difference between the average and the median in this case amounts to almost two months!). Thus, Simonton and his collaborators first compare medians, but at a certain moment they calculate the average of medians recorded by others to their own median. One need not be a statistician to understand how this maneuver can create serious distortions of reality.

To summarize this cursory analysis (in which there are many flaws in the research I have not discussed), it is also worth pointing out that the authors did not conduct any statistical analysis that would determine the extent of the effect explained by the influences of psychotherapy, nor the statistical significance level of the achieved results. They only demonstrated that one number is greater than another. This is not quite enough for scientific studies, even for psychology students preparing a Master's thesis.

The fact that their article was published at all can only be explained by recalling that the journal *Psychosomatics*, in which it was printed, did not referee articles submitted to it at the time. Anonymous reviews would not have failed to mention the flaws I have pointed out, along with many other ones.[247]

When I read articles by scrupulous scientists, particularly those based on quasi-experimental plans, the conclusions generally read like this: "the results indicate the potential which the method possesses, but further research is necessary considering the limitations inherent in interpretation of those results." Or "the results should be treated with a large degree of caution." Meanwhile, the statement, repeated like a mantra, that "complementing traditional treatment with appropriate psychotherapeutic intervention leads on average to a twice-longer survival time for patients," is simply a marketing lie told in the expectation that it will never be demasked. However, substituting the entirely false statement "twice as many patients cured" with the euphemism "the statement is not exact" may be witty, but it fails to amuse when addressed to people with a death sentence. Unless it is assumed with no little cynicism – and correctly, we may add – that they are unlikely to ask for their money back...

Many supporters of the Simonton method eagerly invoke the aforementioned research. None of them, or at least none of those I have encountered, bother to cite the key statement in assessing the effectiveness of the therapy issued in 1981 by the American Cancer Society (ACS) and published in "A Cancer Journal for Clinicians," which, at least through June 2015, is held to be an accurate reflection of the current state of knowledge. As the first sentence already informs us: "After careful study of the literature and other information available to it, the American Cancer Society does not have evidence that treatment with O. Carl Simonton's psychotherapy method results in objective benefit in the treatment of cancer in human beings."[248]

Further fragments of the statement contain a description of the nature of the methods applied at the Simonton clinic, as well as of the therapists' work and analysis of their publications. We also find out that in July 1980, the ACS wrote a letter to Dr. Simonton with a request for information and published materials concerning his methods. The response penned by Dr. Simonton's wife in October of that same year contains a declaration that the center has just begun preparing a five-year cycle of studies designed to better evaluate the results of their work. Since that time – over thirty years ago – the ACS has yet to receive any information about the realization of those plans.

In the Spring of 1981, independently of each other, consultants from the Psychiatry Departments of two leading New York-based medical centers, the Memorial Sloan-Kettering Cancer Center and Mount Sinai Hospital, conducted an analysis of available published materials and video recordings on psychotherapy conducted by Simonton's center. Their conclusions were almost identical. Their statements contain both five positive and five negative assessments. Let us begin with the positive ones, which make for very educational reading.

The first thing that our attention is drawn to is the contribution of Simonton's therapy, which consists in the exhortation to "do something, anything" with cancer, and encourages efforts by the patient to exert control over the situation. The second benefit that the authors of the statement point out is the promotion of relaxation, which can lead to a reduction in anxiety and a temporary improvement of mood. The third positive aspect was counteracting the feeling of helplessness. Fourth among those noted by ASC was the fact that Dr. Simonton does not recommend interrupting the standard recommended medical treatment (e.g. chemo- or radiotherapy). Finally, the authors point out that the emphasis on forming a positive, active attitude in the Simonton method can assist patients in better adapting to the situation.

In relating these conclusions, I feel obliged to comment. Could it possibly be the case that the flood of cargo cult science and quacks enriching themselves on the misfortune of others has led to the mere fact of not consciously doing any harm being considered a virtue?! If I spent a second dreaming up a therapy for cancer patients consisting in watching television for 12 hours a day, but without discontinuing conventional therapy, would my ridiculous recommendations now enjoy the respect of consultants from the ACS?! Is the fact that we do not kill, steal, or rape a virtue? Or perhaps the immense pressure for tolerance of everything that is different, strange, irrational, and sick has led us to completely abandon our faculties?

Among the negative aspects of Simonton's therapy, the experts of ACS indicated both the absence of empirical proof and of scientific bases for statements as to its effectiveness. The second significant reservation concerned the erroneous assumptions made by Simonton, including those regarding the personal responsibility of the patient for the development of the cancer, and then about his participation in the treatment process. In the opinion of the report's authors, there is no evidence to back such a claim. There is also no evidence for what is highlighted in the next statement, that is, the assumption that less stress allows the body to deal better with cancer, or that it at least slows the rate of growth of cancer cells. The fourth point stresses the absence of proof for the statement that using imagination is an effective means of influencing the process of the emergence and development of cancer. The list of charges against the therapy concludes with one about the potential risk to patients associated with making them feel guilty, with excessive dependency on the therapy, and with stopping conventional therapy in spite of the advice of Simonton.

That last risk is of such significance considering that there are various healers employing positive thinking and other similar methods who enjoy incomparably better marketing than traditional medicine and have perfected techniques of persuasion. This is illustrated by a situation that occurred during one particularly popular television program. When the host, the famous Oprah Winfrey, went on one of her shows and advertised the book *The Secret* by Rhonda Byrne, herself a promotor of healing through positive thinking, one woman suffering from breast cancer decided to cease her conventional treatment. In the next episode Oprah did go on to warn strenuously against such practices,[249] but it is a well-known fact that many people can be taken in by marketing promises that absolutely everything depends on our desires and our will. This can easily be learned from studies. For example, among women diagnosed with breast cancer,[250] ovular cancer,[251] and ovarian cancer[252] who have survived at least two years from the moment of their cancer diagnosis, between 42% and 63% of them were convinced that stress was the source of their illness. In other studies it turned out that from 60% to as much as 94% of women believe that they do not get ill because they have a positive attitude towards life! In all of the aforementioned studies there were far more women convinced that their cancer was the result of stress rather than those who acknowledged the role of a range of factors such as genetic determinants and environmental factors.

Let us return, however, to the statement by the ACS and recall its final conclusion: "In summary, the consensus of the consultants at Memorial Sloan-Kettering Cancer Center and at Mt. Sinai Medical Center (see Evaluation), who are psychiatrists and psychologists, was that although in its more positive aspects the Simonton technique may increase patient comfort and ability to deal with cancer, there is no scientific evidence that psychological and psychosomatic factors will alter the course of the disease."[253]

Is it possible that for thirty years the supporters and promotors of the Simonton method have failed to come across this statement? Is it possible that people with a medical education are unfamiliar with it? And is it conceivable that Dr. Mariusz Wirga, who lives and practices in the United States, eager in his support for the Simonton idea both there and in Poland, has not read the text? I have very carefully searched both scientific databases and the Internet to find a comment, response, or discussion involving the declaration at hand. My efforts were in vain, as were my attempts at locating any reference at all to it on a website promoting the Simonton method. Are these people so full of themselves that their private experiences lead them to reject the collective experience and output of the most outstanding experts in cancer treatment? Do they believe in a conspiracy among doctors and pharmaceutical companies acting to keep effective treatment methods out of people's hands in order to continue making money (such explanations are easy to find)? Or perhaps they are simply acting cynically, and they believe that fairy tales of the miraculous power of the mind and desires of people facing death constitute a good way to make a buck?

It is not my job to answer these questions; I prefer to present facts that allow everyone to make their own judgments. Here they are. Several meta-analyses have shown that, contrary to common belief, there is no association between emotions, stress, and cancer rates.[254] Scientists have also failed to identify a link between positive attitudes, emotional states, and cancer survival.[255] Particularly interesting are the results of studies by James Coyne and his coworkers who spent nine years observing 1,093 patients with benign cancer of the neck and head. As it turned out, the attitudes displayed by the patients – from a total loss of hope to boundless optimism – did not make any difference in their life expectancy. The most fatalistic-minded did not live a day longer or shorter than the incorrigible optimists.[256] Similar results were recorded in Australia from a sample of 708 women diagnosed with breast cancer. The study lasted for eight years and demonstrated no connection between negative emotions, depression, anxiety, fear, rage, and pessimistic attitudes on the one hand, and expected survival rates on the other.[257]

Interestingly, in proper studies we can even observe the opposite effect. The association so eagerly pushed by Simonton and other healers pretending to the title of psycho-oncologists between the intensity of stress experienced and susceptibility to cancer has not only not been proven (as I have already written), but in fact the opposite association has been shown, and that in a massive research sample. The study I am referring to encompassed 37,562 nurses in the United States, and lasted for eight years (1992–2000). Their results led to the statement that the risk of breast cancer is 17% lower among women experiencing relatively greater stress in their work! Danish researchers have generated an even more striking result. While the sample size was indeed somewhat smaller (just under 7,000 women), the study lasted twice as long. It turned out that the group reporting a high level of stress in the workplace demonstrated a 40% lower risk of developing breast cancer.[258]

Equally surprising are the results of studies on the effects of encouraging patients to find the positive sides of their illness. In one recently published report it turned out that women who were taught to regard their cancer diagnosis as a beneficial situation declared lower quality of life, as well as worse psychological functioning, when compared to those who saw no benefits in their disease.[259] Other research has shown that patients with breast cancer perceive intense efforts at encouraging them to see the positives in their illness as utterly inadequate to the situation they have found themselves in. They practically always interpreted such pressure as an attempt at minimizing their suffering.[260]

These effects are explained by Barbara Ehrenreich, author of *Smile or Die. How Positive Thinking Fooled America and the World*, and these are not purely theoretical considerations; she herself went through breast cancer and experienced all of the feelings and sufferings associated with it.

> It could be argued that positive thinking can't hurt, that it might be a blessing to the sorely afflicted. Who would begrudge the optimism of a dying person who clings to the hope of a last-minute remission? Or of a bald and nauseated chemotherapy patient who imagines that the cancer experience will end up giving her a more fulfilling life. Unable to actually help cure the disease, psychologists looked for ways to increase such positive feelings about cancer. ...
> If you can't count on recovering, you should at least come to see your cancer as a positive experience. ...

But rather than providing emotional sustenance, the sugar coating of cancer can exact a dreadful cost. First, it requires the denial of understandable feelings of anger and fear, all of which must be buried under a cosmetic layer of cheer. This is a great convenience for health workers and even friends of the afflicted, who might prefer fake cheer to complaining, but it is not so easy on the afflicted.[261]

The tyranny of positive thinking has also been resisted by the creator of psycho-oncology. Jimmie Holland says unequivocally that encouraging patients to think positively is not only an extra burden imposed on them in a difficult situation, but also deprives them of vital social support. Many ill people adopt the understanding of positive thinking supporters, and begin to feel a sense of guilt over their illness.[262]

The comparison of metaanalyses, cohort studies, and reviews of research with the previously analyzed and unreviewed study by Simonton, which 225 patients took part in, is like placing a division equipped with cannons, shells, and grenades against a soldier carrying a homemade pipe bomb. Indeed, from time to time a brave soul appears and makes an attempt to inject similar "results" and unique conclusions into the scientific discourse, but in the face of failed replications he quickly retreats. One such example is that of a study conducted by David Spiegel and his collaborators, which was supposed to be proof that women with metastasized breast cancer who participated in support groups lived twice as long as those who didn't attend such meetings.[263] Unfortunately, for almost twenty years, nobody – including Spiegel - has succeeded in repeating that result.[264]

Pretenders to the title of psycho-oncologists remain undeterred. The previously mentioned Mariusz Wirga, probably the most well-respected Polish representative of the discipline, can boast only one publication in the EBSCO scientific database, and that is a non-empirical, theoretical piece. However, on Polish web pages promoting the Simonton program, in the "Recommended Literature" section we can easily find three articles of his authorship that have appeared in a popular magazine. Another two, by other authors, were also printed in the entertainment press.[265] Well, patients with a death sentence most certainly have neither the time nor the desire to seek out evidence of their therapy's effectiveness in scientific journals. Every second counts for them. Unfortunately, sometimes just a second is enough for a recommendation to replace proof, and for marketing to shield our eyes from reality. No matter what else we might say about them, the marketing skills of pseudo-oncologists cannot be denied.

Their skills are also evidenced in the manner how they attempt to develop associations with tried and tested scientific methods. For some time, Simonton and his imitators have included in their therapeutic program elements they refer to as Maultsby Rational Behavior Therapy. The therapy is based on the Five Rules of Rational Thinking, the first of which states that "Rational Thinking is based on obvious facts."[266] Hmm. "Obvious facts" is already a concept up for discussion. Perhaps this is only a message for patients, and doesn't concern therapists? Clearly, in certain situations the strength taken from the Chinese healer sitting in the yoga position on the astral plane and given to us by Silva becomes "rational."

This, however, is not the end of efforts at rendering things more scientific. Alongside the name Simonton ("the creator of psycho-oncology"), we come with increasing frequency across the term "behavioral-cognitive therapy." As we wrote in *Psychology Gone Wrong*, this is a therapy that achieves the best results in effectiveness studies. It is the most thoroughly tested method, and this makes it undoubtedly the most effective method in psychotherapy for those suffering from cancer. Jadwiga Rakowska provides a summary of the large body of literature on the subject: "In the treatment of cancer, cognitive-behavioral therapy, encompassing training in relaxation, stress-management and problem-solving techniques, is an effective method in achieving better mood, reducing psychological suffering and improving perception of quality of life."[267]

But there is nothing here about living longer or quality of death! (As an aside, I have long wondered how it is that supporters of Simonton measure their so heavily promoted quality of death?). And perhaps this is a lot? After all, it is the most effective therapy in bringing relief to patients suffering from cancer. It is most definitely far ahead of others, particularly in studies identifying evidence-based practice. That said, other than some superficial similarities, it has nothing in common with Simonton therapy. Yet associations look good and pick the pockets of the clothes worn by the ignorant...

Now we come to arrive at the real role of psychotherapy in helping cancer patients. This therapy can improve one's mood, reduce psychological suffering, have a beneficial impact on perception of quality of life, and it may even reduce the feeling of pain. Unfortunately, there is as of yet no proof that it impacts the development of cancer cells, or that it is associated with longer life expectancy.[268] If such dependencies exist, they should be examined immediately, their mechanisms understood, and psychotherapy methods perfected. It will be the task of researchers and administrators in the scientific community to assess the potential for success of such studies, and to determine whether it is worth spending money on them rather than on medical means of extending and improving the quality of life.

We are all under a duty to accompany our loved ones through the most difficult moments in their life's journey, to comfort them and to give them hope. If we are incapable of doing this ourselves, or if the task overwhelms us, there is no reason to avoid seeking the help of professionals, including psychotherapists. The quality of life experienced by loved ones at death's door becomes priceless in such moments. However, each of us should decide independently and in accordance with our own conscience where the giving of hope ends, and where falsehoods begin; we must also decide whether we are prepared to lie consciously. During TV and radio programs promoting my books, I had a number of conversations with the loved ones of people suffering a terminal illness. My interlocutors said that they remain unable to cope with the fact that, up to the last, they deceived those closest to their heart, with the very best of intentions. Looking at it from the perspective of time gone by, they are certain that they would not hesitate to tell the truth, which would allow for their last moments with their beloved to be experienced with full awareness and the gravity the situation deserves.

A deadly illness and its terminus are extreme situations; too difficult to be judged from the perspective of an outsider. However, when the process is joined by professionals, not only can we, but we should assess what they are doing. The effectiveness of psychotherapy according to the methods of O. Carl Simonton has, for almost forty years, not been verified in accordance with the standards applicable in science and medicine. The majority of trustworthy studies indicates that the descriptions of effectiveness in advertisements for the method are false. The sources invoked by its supporters do not account either for the position of the American Cancer Society, nor other critical studies. I shall allow you to draw your own conclusions as to the morality of the situation.

---

[231] As I have been told by oncologists, there is presently a strong tendency in the case of cancer to speak of a chronic illness rather than a terminal one. In some cases, this is justified. In others, it is nothing more than a synonym for the process of dying slowly.

[232] R. Piana, "Jimmie Holland, Central Founder of Psycho-Oncology," *Oncology News International 18*, (2009): 2.

[233] I conducted my search in June 2015 Considering the continually changing nature of content on the Internet, the results themselves are equally changing. Undoubtedly, the raw numerical data will change quicker than proportions. This is confirmed in the very similar proportions that I recorded conducting identical searches, first in December 2010, and again in November 2011.

[234] O. C. Simonton, R. Henson, and B. Hampton, *Powrót do Zdrowia* (Łódź: Ravi, 1996).

[235] http://www.simontoncenter.com/founder.asp

[236] http://www.thesimontondocumentary.org/about_dr_simonton

[237] Indeed, Louise Hay, known in the USA as the "Queen of the New Age" and author of self-help books – including on the subject of cancer treatment – claims to have been diagnosed in 1977 or 1978 with cervical cancer. The date is not definite, as Hay claims to be unable to recall when, exactly, it happened. She says that she refused medical therapy and treated herself, with the help of reflexology, diets, cleansing enemas and acts of (self)forgiveness. The sole attestment of her recovery is her own words. There are no other witness, nor any medical documentation. When asked for details, she invariably replies "That was years ago." See: M. Openheimer, "The Queen of the New Age," *The New York Times*, (May 4, 2008).

[238] http://en.wikipedia.org/wiki/Alexander_Everett

[239] D. Hunt, and T. A. McMahon, *America, the Sorcerer's New Apprentice: The Rise of New Age Shamanism* (Eugene, OR: Harvest House, 1988), 64.

[240] http://www.silva.alpha.pl/
[241] http://www.silvamethodlife.com/accelerate-the-natural-healing-process/
[242] http://www.simonton.pl/node/1
[243] http://aureliadembinska.pl/?page_id=46
[244] http://wroclaw.gazeta.pl/wroclaw/1,79448,5735986,Powstala_pierwsza_poradnia_psychoonkologiczna.html
[245] Private correspondence.
[246] O. C. Simonton, S. Matthews-Simonton, and T. F. Sparks, "Psychological Intervention in the Treatment of Cancer," *Psychosomatics, 21*, (1980): 226-227, 231-233.
[247] The second publication Wirga recommended to me was equally poor, and had also not been reviewed prior to publication: O. C. Simonton, and S. Matthews-Simonton, "Cancer and Stress: Counselling the Cancer Patient," *The Medical Journal of Australia 27*, (1981): 679, 682-683.
[248] American Cancer Society, "Unproven Methods of Cancer Management. O. Carl Simonton, M.D.," *A Cancer Journal for Clinicians 32*, (1982): 58-66.
[249] As cited in: S. O. Lilienfeld, S. J. Lynn, J. Ruscio, and B. L. Beyerstein, *50 Great Myths of Popular Psychology* (Chichester, UK: Wiley-Blackwell, 2010), 217.
[250] D. E. Stewart, A. M. Cheung, S. Duff, F. Wong, M. McQuestion, T. Cheng et al., "Attributions of Cause and Recurrence in Long-Term Breast Cancer Survivors," *Psychooncology, 10*, (2007): 179-183.
[251] D. E. Stewart, S. Duff, F. Wong, C. Melancon, and A. M. Cheung, "The Views of Ovarian Cancer Survivors on its Cause, Prevention, and Recurrence," *Medscape Womens Health, 6*, (2001): 5.
[252] C. Costanzo, S. K. Lutgendorf, S. L. Bradley, S. L. Rose, and B. Anderson, "Cancer Attributions, Distress, and Health Practices Among Gynecologic Cancer Survivors," *Psychosomatic Medicine, 67*, (2005): 972-980.
[253] American Cancer Society, *Unproven Methods*.
[254] P. N. Butow, J. E. Hiller, M. A. Price, S. V. Thackway, A. Kricker, and C. C. Tennant, "Epidemiological Evidence for a Relationship Between Life Events, Coping Style, and Personality Factors in the Development of Breast Cancer," *Journal of Psychosomatic Research, 49*, (2000): 169-181; S. F. A. Duijts, M. P. A. Zeegers, and B. V. Borne, "The Association Between Stressful Life Events and Breast Cancer Risk: A Meta-Analysis," *International Journal of Cancer, 107*, (2003): 1023-1029; M. Petticrew, J. M. Fraser, and M. F. Regan, "Adverse Life-Events and Risk of Breast Cancer: A Meta-Analysis," *British Journal of Health Psychology, 4*, (1999): 1-17.
[255] B. L. Beyerstein, W. I. Sampson, Z. Stojanovic, and J. Handel, "Can Mind Conquer Cancer?" In *Tall Tales About the Mind and Brain: Separating Fact from Fiction*, ed. S. D. Sala (Oxford: Oxford University Press, 2007): 440-460.
[256] J. C. Coyne, T. F. Pajak, J. Harris, A. Konski, B. Movsas, K. Ang et al., "Emotional Well-Being does not Predict Survival in Head and Neck Cancer Patients: A Radiation Therapy Oncology Group Study," *Cancer, 110*, (2007): 2568-2575.
[257] K. A. Philips, *Psychosocial Factors and Survival of Young Women with Breast Cancer*, Presentation at Annual Meeting of the American Society of Clinical Oncology, Chicago, (June, 2008).
[258] N. R. Nielsen, Z. F. Zhang, T. S. Kristensen, B. Netterstrom, P. Schnor, and M. Gronbaek, "Self-Reported Stress and Risk of Breast Cancer: Prospective Cohort Study," *British Medical Journal 331*, (2005): 548.

[259] M. Dittman, "Benefit Findings Doesn't Always Mean Improved Lives for Breast Cancer Patients," *APAOnline*, (February, 2004).
[260] H. Tennet, and G. Affleck, "Benefit Finding and Benefit Remainding," in *Handbook of Positive Psychology*, ed. C. R. Snyder, and S. J. Lopez (New York: Oxford University Press, 2002), 584-597.
[261] B. Ehrenreich, *Smile or Die. How Positive Thinking Fooled America and the World* (London: Granta Publications, 2009), 40-41.
[262] J. Holland, and S. Lewis, *The Human Side of Cancer, Living with Hope, Coping with Uncertainty* (New York: HarperCollins, 2000), 13-25.
[263] D. Spiegel, J. R. Bloom, and E. Gottheil, "Effects of Psychosocial Treatment on Survival of Patients with Metastatic Breast Cancer," *Lancet 2*, (1989): 888-891.
[264] Beyerstein et al., "Can Mind Conquer."
[265] http://www.simonton.pl/taxonomy/term/6
[266] M. C. Maultsby, *Rational Behavioral Therapy* (Appleton, WI: Rational Self-Help Aids/I'ACT, 1990), 8.
[267] J.M. Rakowska, *Skuteczność Psychoterapii* (Warszawa: Wydawnictwo Naukowe SCHOLAR, 2005), 157.
[268] J. C. Coyne, M. Stefanek, and S. C. Palmer, "Psychotherapy and Survival in Cancer: The Conflict Between Hope and Evidence," *Psychological Bulletin 133*, (2007): 367-394.

Chapter 9: The Disaster Industry and Trauma Tourism: The Harmful Effects of the Psychological Debriefing

It was a chilly and cloudy day. Nonetheless, we got up early, prepared sandwiches and a thermos of tea, and got our backpacks ready. A one-hour drive brought us to the foot of the mountain whose peak we were anxious to reach. The lift, which could have made our route less painful, turned out to be closed due to the strong wind that day, thus giving us a vast extra distance to walk. With no hesitation, however, we took up the challenge and started the climb. Four hours later and there we were, standing on the summit and admiring the magnificent view. Although it was still September, the weather up there resembled winter conditions. For our descent we took a longer but easier way. We got to our car exhausted, chilled to the marrow and perfectly satisfied. Reveling in the beauty of the moment and the raw charm of the surrounding mountains, we drove back to the hotel without switching the radio on. Everyone was yearning for a warm meal, a bone-soothing shower, and rest.

Although it was nearly fourteen years ago, I remember this day as if it occurred yesterday. Many people from our culture who were sufficiently mature at the time remember this day in every detail. It is well understood that in our minds we tend to see the bad as being more powerful than the good. And this very day was inconceivably bad: it was September 11th, 2001.

If most of us, as distant observers, can recall that day with rare clarity, how is it remembered by those who were direct witnesses or even victims of the unthinkable events that unfolded before the eyes of the entire world? Are they able to bear their memories of this horrible day? Do they relive it in their dreams? How often? To what extent do those memories affect their everyday functioning? It seems that most of us have enough empathy to at least hazard a guess as to what answers would be given to all these questions. With a high degree of certainty it can be presumed that those people may be struggling with severe anxiety, feelings of helplessness, nightmares, flashbacks or other symptoms of post-traumatic stress disorder (PTSD).

For many this tragedy goes beyond our understanding. What the people affected by the tragedy saw or experienced cannot be related or described by me, a person who spent that day undisturbed, far away in the tranquility of the mountains. For those of you who obviously cannot remember that far back, let me render the general feeling of the day by quoting excerpts from eyewitness accounts:

> Among the papers and melted steel fragments fluttering to the ground, I notice that some debris was falling distinctly differently. These weren't parts of the building that were falling. These were people, jumping from the windows, their

bodies tumbling in rapid descent from the eightieth floor. I noticed about ten such falls, morbidly capturing three of them on tape.[269]

Can you imagine the smoke and the heat. People don't want to burn to death. I saw about 15 fall or jump. First there was one, about two or three minutes later there was another. I had tears in my eyes. I felt the emotions. I'm shocked.
...
There were so many running down the stairs; running over each other and screaming and pushing and trying to get out. And that was before the tower came down. That was before I learned that my traders and friends were still up there, on the 92nd storey, and I don't reckon I'll ever see them again.
...
I was trying to calm a woman down and tell her the worst was over when a huge piece of debris fell through the ceiling. There was a violent eruption and the building was rumbling. That was another test of wills. Then all the lights went out and the building was blacked out. We just couldn't see a thing and all the time there was water coming like a flood behind us, and smoke and fires.
We got down to the outside and it was an apocalypse. I can't imagine what an atomic bomb is like but maybe this it. We thought we were done then but we had to work our way through piles of debris and all the trees were stripped in a four block radius. Now I need to see if I can find some of my people. I've only got four of them so far.[270]

On this very day and many others to come, thousands of psychologists, psychotherapists and social workers already knew what emotional upheaval the survivors would likely go through afterwards. Driven by admirable humane impulses, they voluntarily decided to assist and give mental and life support to the victims of the World Trade Center attack. In the face of the catastrophe, their aim was simple – to prevent people from experiencing further psychological consequences from the trauma, namely PTSD. To do this, they used the psychological debriefing model (PD), also known as critical incident stress debriefing (CISD) or critical incident stress management (CISM). It consists of a series of interviews that are meant to allow individuals to directly confront a tragic event and share their emotions with the counselor. By doing so, respondents may better structure their memories of the event. The PD sessions often provide a chance to talk about the trauma with others who were also involved. A PD usually takes place immediately after or within the first three days of such a crisis. The method is applied to provide psychological aid to people who were hurt, saw others hurt or killed, or served as first responders.

On September 11th, 2001, most psychologists helping survivors of the terrorist attack used psychological debriefing (PD) to protect them from PTSD. Did they succeed in achieving such a commendable goal? The bitter and sobering answer is NO. They could not have because what they were trying to do in the aftermath of the WTC disaster was worth nothing more than most of the cargo rituals that the Pacific Ocean islanders had acted out.

What is even worse, these rituals could have been harmful for many of the survivors and witnesses of the tragedy. Taking note of their noble urge to help others, can the psychological support volunteers be excused? Again, the answer is NO. It is every professional's responsibility to acquire knowledge of methods that can be effectively used in practice. At that time they ought to have been alert to two factors that later had a profound impact on their efforts: the lack of scientific value of PD, and the healing power of human resilience, which keeps most of us unaffected by such disorders as PTSD.

The origins of the PD method date back to the beginning of the 1980s and its author, Jeffrey Mitchell, who was a firefighter and a paramedic. Here is how Richard Gist, a consulting psychologist at the Missouri Fire Department and Joseph Woodall, the director of the Public Safety Administration at Grand Canyon University, commented on the then social background and vital needs which prompted the emergence of the method:

> Leaders among fire service and emergency medical providers were beginning at that juncture to pay increasing heed to the impact of major events on the wellbeing of their personnel. Certain experiences seemed to confront even seasoned personnel with images and perceptions that were unusually difficult to reconcile. Almost everyone could recall some such experience in his or her own career, and nearly everyone could recall someone whose career seemed to unravel after exposure to some particularly poignant or gruesome experience. If we could help with these impacts, the reasoning went, it was surely the right thing to do.
> …
> Those veteran to rookie chats had long been a part of fire service tradition, but were never formally prescribed and were usually done very quietly and very privately; the support that one received depended pretty much on "the luck of the draw" respecting officers and comrades. The notion that some step should occur to ensure that they were done whenever necessary, done systematically, and done effectively seemed only reasonable; that they could be initiated through organizational protocol and done in a structured group setting seemed not just plausible but inherently sensible. "Critical Incident Stress Debriefing" (CISD) was born.[271]

Somewhere along the way the belief had spread that it was always better to get your emotions out than bottle things up, so debriefing fit right in with that belief. Soon it appeared that this concept was also created at a very opportune moment, and it became very popular among members of many crisis centers. Its basic and very attractive idea of giving psychological support to first responders was not the only underlying cause of its instant success, however. It was also carefully packaged and cleverly marketed by its originator and proponents. A primary characteristic of the CISD movement has been its staunchly argued and often repeated claim that it is a "scientifically tested and proven" intervention scheme created through systematic research.[272]

For the first decade from its inception, the method went virtually unmentioned in the mainstream psychological literature. It rose to prominence in the less critical venues of trade journals, trade shows and through a range of psychological training events. But with the passage of time, a large body of research has been compiled, and a vast base of empirical data has been accumulated.

The first critical study that I tracked down was done by an Australian psychiatrist, Alexander McFarlane, as far back as in 1988. Over a 25-month period he examined a group of 469 firefighters exposed to a bushfire disaster. He noticed that six months after the disaster, one of the best predictive risk factors that enhanced the likelihood of a person developing PTSD was having gone through the psychological debriefing procedure. Moreover, based on the data, he concluded that exposure to extreme trauma is necessary but not sufficient to explain the onset and pattern of posttraumatic morbidity.[273]

One year later, two Norwegian psychiatrists, Karsten Hytten and Anders Hasle, examined 58 non-professional fire fighters who had been engaged in a hotel fire rescue operation. Together with 57 professional fire fighters, they had participated in rescuing hotel guests confined for as long as three hours in a burning 12-story hotel building. Fourteen people died as a result of the fire while 114 guests survived. Almost half of the non-professional fire fighters reported that the disaster experience was the worst they had ever undergone. Even so, 80% thought that they had coped with the job well to fairly well, and for as many as 66% the rescue action represented a positive episode to them in retrospect. The study did not confirm the protective effect of PD on the risk of PTSD.[274]

In 1991, two subsequent studies, both examining groups of UK police officers, were published. None validated (the claim of) the protective effect of PD, and both failed to demonstrate high levels of post-traumatic distress or psychiatric morbidity among officers exposed to high level of distress.[275]

Still, the evidenced lack of protective effect of PD turned out to be only the tip of the iceberg. Later studies showed additionally harmful effects of PD. The Australian researchers Jean Griffiths and Rodney Watts examined the relationship between stress debriefing and stress symptoms among 288 emergency personnel involved in one of the worst Australian traffic accident rescue missions. The personnel who had attended debriefing sessions reported significantly higher levels of symptoms at 12 months than those who had not received the intervention. Furthermore, no relationship was found between the perceived helpfulness of psychological debriefing and the appearance of symptoms.[276]

The effectiveness of PD was likewise not confirmed in British soldiers whose duties included the handling and identification of dead bodies of allied and enemy soldiers during the Gulf War. The soldiers who had been debriefed did not have lower levels of morbidity than those who had not been debriefed.[277]

Year after year more data emerged. In 1994, the data mass was sufficient for investigators to publish the first reviews. One of them was written by the British scientists Jonathan Bisson and Martin Deahl. They came to the conclusion that the hypothesis that PD decreased psychological sequelae had not been adequately tested. In their opinion, the data available at that stage suggested that, at best, PD afforded some protection against later sequelae and, at worst, makes no difference.[278]

In the same year, Donald Meichenbaum from the University of Waterloo published *Clinical Handbook/Practical Therapist Manual for Assessing and Treating Adults with Posttraumatic Stress Disorder*, in which he described and criticized PD by exposing situations where it created more harm than good.[279]

One year later, two thorough reviews of the literature on psychic trauma were published. The first, written by American specialists in fire science Richard Gist and Joseph Woodall, discussed models of addressing stress disorders, and argued for a reasoned, ethical and effective approach to the problem. They did not find reliable empirical evidence indicating any demonstrable preventative effect of PD.[280] In the second study, the authors also tried to answer the question of whether protective programs such as PD ward off PTSD. Again, their answer was very pessimistic.[281]

In 1996, Michael Hobbs, Richard Mayou, Beverly Harrison and Peter Worlock from Oxford Hospital published the results of their research on the efficacy of psychological debriefing in individuals who had been involved in road traffic accidents. The study demonstrated no differences between the group of road accidents victims (the intervention group) and the control group.[282] Both groups were reexamined after three years. This time the study reported that those who had received the PD intervention showed higher levels of PTSD symptoms, increased travel anxiety, and reported more frequent physical complaints than the control group. Despite the psychological debriefing having been offered to them, after three years the patients who had experienced significant post-traumatic stress did not show any improvement, with their assessment scores being much worse as compared with the control group.[283]

That same year, an Australian clinical health psychologist at the University of Queensland, Justin Kenardy, along with his colleagues, published results of their research in which they examined the efficacy of PD offered to individuals who had themselves provided psychological support after an earthquake in Newcastle in 1989. Two years after having received the intervention, as compared to 133 non-debriefed helpers, 65 debriefed helpers who were examined showed greater adaptive difficulties and increased symptoms of acute stress disorder.[284]

The second half of the decade saw a great abundance of research, articles and reviews devoted to PD, its effectiveness and harmful effects.[285] As it is impossible to elaborate on all of them, let me mention just a few. First, there was a study published by Gist and his colleagues in 1998. They investigated the structure and efficacy of PD following the crash of an airliner in which 112 of 296 passengers died. The findings of that study, including an almost complete sample of carrier firefighters engaged in body recovery and related operations, showed neither a clinically significant impact on the personnel at two years post-incident, nor evidence of superior resolution for debriefed responders versus those who declined. Additionally, evident in the study was a slight but statistically significant trend toward worsening in resolution indices for those accepting PD, and a clear preference for informal sources of support and assistance that correlated strongly with effective resolution.[286]

Other important articles published in the second half of the first decade included systematic research reviews on the efficacy of a single debriefing session carried out by Suzanna Rose and Jonathan Bisson, psychologists from the University of London. They were able to find only six randomized controlled trials. Two studies associated the intervention with a positive outcome, two demonstrated no difference on outcome between intervention and non-intervention groups and two others showed some negative outcomes in the intervention group.[287] Rose did one more research review in 2000. The only conclusion from that analysis was that the use of PD was at best contentious.[288]

The most alarming doubts as to the efficacy of debriefing came with the Cochrane Review and its analysis of the most recent fifteen years of research into psychological debriefing. The results demonstrated that debriefing interventions were found to be either harmful or ineffectual. The first review was done in 1997 and published in 1998. It has been regularly updated since then.[289] Its 2000 version offered the following, unequivocal conclusion: "There is no current evidence that psychological debriefing is a useful treatment of posttraumatic stress disorder after traumatic incidents. Compulsory debriefing of victims of trauma should cease."[290]

Of course, PD is not the only unsupported procedure that was designed and developed with a view to helping traumatized people. Recently several so-called "power therapies" have been also conceived. Their collective name draws on the claim that such treatments work much more efficiently than extant interventions for anxiety disorders and exert their effects through processes that are distinctly different from psychological insight or learning. The power therapies include eye movement desensitization and reprocessing, thought field therapy, emotional freedom therapy, traumatic incident reduction and visual/kinesthetic dissociation. I will limit myself to indicating their names only, as these approaches are much less popular among trauma helpers and even much less scientifically proven. A curious reader will find many sources to dig deeper into the subject.[291]

So far I have presented just some of the evidence with regard to PD available before the WTC attack. Much of this was accessible long before

9/11. Every practicing professional at that time should have known it. It is also worth mentioning that, by then, scientific research had revealed a number of efficacious treatment procedures for the amelioration of PTSD symptoms. Most were based on cognitive-behavioral theory and involved strategies that had withstood rigorous scientific scrutiny. So why is it that most psychologists and therapists applied the dubious PD approach to the WTC survivors on a mass scale that had no precedent in history? The answer is yours to discover.

After the WTC attack, critical data showing the lack of efficacy of PD or even exposing its iatrogenic role were continuously accumulated. However, the problem was only brought to light as late as on the tenth anniversary of the WTC disaster. In 2011, it came out through a special issue of APA's flagship journal, *American Psychologist*, with a dozen peer-reviewed articles. The issue, also called a report, offered numerous examples of how psychology was helping people in understanding and coping with 9/11's enduring impacts. It also explored ways in which psychological science had aided us in becoming aware of the roots of terrorism and how further attacks could have been prevented. Even before this publication, the mass media reacted to its content. One of the first prominent journals to make a comment was *The New York Times*. What they wrote seemed to be shockingly surprising for everybody: "Experts greatly overestimated the number of people in New York who would suffer lasting emotional distress. Therapists rushed in to soothe victims using methods that later proved to be harmful for some. ... But researchers later discovered that the standard approach at that time, in which the therapist urges a distressed person to talk through the experience and emotions, backfires for many people. They plunge even deeper into anxiety and depression when forced to relieve mayhem."[292]

Later? Can it be true?? What about all those studies published before 2001 that I have cited in this chapter?

Roxane Cohen Silver, a psychologist from the University of California, Irvine, and the author of "An Introduction to '9/11: Ten Years Later,'" while interviewed by *The New York Times*, seemed to be surprised as well. "'You have to understand,' she said, 'that before 9/11 we didn't have any good way to estimate the response to something like this other than – well, estimates based on earthquakes and other trauma.'"[293]

In the research that has been presented earlier in this chapter, carried out before 2001, several thousand people were examined who had been affected by a number of various disasters. Had there really not been any opportunity to evaluate people's reactions to trauma and ways of providing help in coping with post-incident symptoms?

Three days later, on July 31st, a well-known British tabloid – the *Daily Mail* – published an article with a distinctively dramatic title: "Therapy can Drive you Mad, Finds Study on Counseling Given to 9/11 Survivors." John Stevens, the author of this article, argued that:

Therapy can exacerbate trauma and make things worse according to a study looking at the counseling given to New Yorkers in the aftermath of 9/11.

The report, to be published in the journal *American Psychologist* next month to coincide with the 10th anniversary of the September 11 attacks, found that reliving the events was harmful for many survivors.

Mental health professionals flooded the city in a wave of 'trauma tourism' after two planes struck the World Trade Center in 2001 according to the report.

But the main psychological benefits were felt by the psychologists rather than the patients, said the study, which said experts greatly over-estimated the number of people who wanted treatment.

'We did a case study in New York and couldn't really tell if people had been helped by the providers – but the providers felt great about it,' Patricia Watson, a co-author of the report who works at the National Centre for Child Traumatic Stress told The New York Times.

'It makes sense; we know that altruism makes people feel better.'

According to the report, therapy centers were set up in the offices of major employers and in fire stations after 9/11.

But for many survivors, the standard procedure at the time of asking them to talk through their experience was not helpful.

Researchers believe that the process can sometimes push people deeper into depression and worsen anxiety.[294]

Overstated as it may have been, again the article reflected well the sense of sudden astonishment among scientists and therapists.

An article in *Psychology Today* took a similar tone. Susan Krauss, its author, describes 9/11 as a priceless lesson learnt by psychologists. "The first lesson from 9/11, then, is that not everyone responds the same way to a disaster. For whatever reasons, those who qualify as 'resilient' will recover on their own. It's those at risk for PTSD, however, who stand to benefit from psychological intervention.[295]

Indeed, it was a powerful lesson, in particular for all who had not done their homework before. Those who liked to go the extra mile could have easily learned it from Alexander McFarlane's study published in 1988 and from many other works published later. A statement from Patricia Watson, co-author of the report, seems to convey much the same feeling of confusion when pondering the effects of PD: "Watson noted that she couldn't tell if the providers helped people or not. The helpers felt better, but did those they were trying to help? They employed the standard of care known at the time of 'psychological debriefing': to urge a distressed individual to talk through the experience and emotions. Unfortunately, this approach has the opposite effect on many people who become more depressed and anxious when reliving the trauma.""[296]

Let me once more make the point that, just as Patricia Watson did, we all had enough opportunities to learn those lessons long before 2001.

Without a doubt, there was a great deal of exaggeration in the media with regard to the report. However, when I review it, some of the sentences and formulations capture my attention more than others. The first sentence of the article entitled "Postdisaster Psychological Intervention Since 9/11" reads: "A wealth of research and experience after 9/11 has led to the development of evidence-based and evidence-informed guidelines and strategies to support the design and implementation of public mental health programs after terrorism and disaster."

As hard as it may be to refute this claim, it begs the question: what were the strategies based on before 9/11? On faith? As I have presented earlier, there was enough evidence to build those strategies on without having to resort to the tradition of useless empty rituals.

The WTC disaster was an unprecedented event in the history of Western civilization. Even though that may be the case, it can also be recognized as an unprecedented failure of psychologists and psychotherapists who, instead of bringing aid and relief to those traumatized, engaged themselves in empty and sometimes harmful rituals of no practical value to anybody. The special issue of *American Psychologist* brought about an atmosphere of total surprise and uncertainty, which was captured by the media in an instant. I am unable to interpret it in any other way than as putting up a smokescreen to hide one of the most disturbing downfalls in the short history of psychology as a science.

Nowadays, the uselessness of PD and early post-trauma interventions is widely and openly commented not only in the specialist literature but even in newspapers and magazines.[297] It seems that it is common knowledge, and any individual who shows even a faint interest in psychology should be aware of it. This is an evidence-based approach to everyone but cargo cult priests. True believers cling tenaciously to their rituals, whether under fire from empirical research or caught in the overwhelming flood of common sense.

At the same time as the report was published, a therapist and an author of the debriefing manual, Debbie Hawker, along with John Durkin, a trainer and consultant to fire, rescue and paramedic services as well as a voluntary debriefer, together wrote an article "To Debrief or not to Debrief our Heroes: that is the Question" in which they claimed: "Several randomized controlled trials tested truncated forms of debriefing in a different population: primary victims of unexpected trauma. These trials, and particularly two in which debriefing appeared to be harmful, led two major reviews to warn practitioners not to offer debriefing. Consequently, many organizations have stopped providing debriefing to employees who face trauma in their routine work."[298]

In their opinion these two studies were seriously flawed. It must have escaped their notice that more of such studies had been done and their results published almost two decades earlier, because they later concluded as follows: "We have been told that the case against debriefing is proven and the debate is closed. We disagree. ... We predict that appropriate psychological debriefing will be shown to have benefits for secondary victims of trauma who have been briefed together and who have worked together through traumatic events. Research into these uses of debriefing should be encouraged and supported."[299]

In response to these assertions, Christian Jarrett, a psychologist and journalist, posed the following question: "Is it time to resurrect post-trauma psychological debriefing for emergency responders and aid workers?" He also discovered that: "Perhaps it's no surprise that post-trauma psychological debriefing is surfacing under new names like 'powerful event group support' and 'trauma risk management.'"[300]

Following the trail marked by Jarret, it is very easy to find a plethora of services designed to support traumatized people.[301] Old wine in new bottles...

In 2013, almost two years after the tenth anniversary of 9/11, Karen Paul from the University of Calgary and Mark van Ommeren at the World Health Organization in Geneva published an article entitled "A Primer on Single Session Therapy and its Potential Application in Humanitarian Situations." Were they familiar with the research results presented in the special issue of the *American Psychologist*? The odds are they were not. One cannot find any reference to those articles in their bibliography. Had they read them, they would have known that: "The research generally confirms that the majority of disaster survivors will not typically require the attention of mental health professionals. Additionally, because resilient individuals already appear to cope effectively on their own, it is possible that global interventions hold no advantage or might even undermine natural coping abilities."[302]

They would have also been aware that: "A number of systematic reviews on early intervention for trauma survivors have concluded that there is very limited evidence for any psychological intervention within the first month following any type of traumatic event."[303]

Instead, they insisted that the following points should be given attention: "As single session services continue to expand in high income countries, careful attention should be given to its potential for adaptation to fit the needs of services in emergency settings in low and middle income countries as well. This is especially true where only one session is feasible. As research continues to explore single session's potential to strengthen mental health services in high income countries, the potential to provide single session services in a manner that strengthens mental health services in emergency situations should also be explored."[304]

Will the authors, by the same token, be also surprised if, ten years from now, a similar disaster happens and the helpers likewise apply their ineffective and potentially harmful procedures? This is very likely, as is the likelihood that they will not be alone. Jeff McGill, an adjunct professor of criminal justice and the author of articles written for PoliceOne.Com will join them. In one of his papers, written in May 2015, he carelessly stated: "Critical incident debriefings conducted by peer supporters is the best practice in place now for addressing the immediate reactions to traumatic stress for law enforcement officers."[305]

Does he not know that the National Fallen Firefighters Foundation — an organization responsible for creating standards and requirements for fire prevention and suppression, training, equipment, codes and standards — removed all recommendation for using CISM from its 2009 standards?[306] Is he unfamiliar with the current WHO recommendations as regards psychological debriefing? "Psychological debriefing should not be used for people exposed recently to a traumatic event as an intervention to reduce the risk of posttraumatic stress, anxiety or depressive symptoms.

Strength of recommendation: STRONG."[307]

The last 30 years of research, and the special issue of the *American Psychologist* in particular, have clearly proven the following, simple claims:
- Most people are resilient enough to withstand a trauma.
- It is quite easy to harm people in trauma, therefore the best form of support given to a traumatized person after the event should be simple and should steer clear of structured psychological interventions.
- The most effective method is to let people find a way to overcome trauma by themselves
- Whether a traumatized person needs more psychological support cannot be determined before the lapse of one month after the trauma.

But there is no power on this earth able to stop cargo cult rituals. They are flexible enough to turn these simple truths to their favor: "'Most people are resilient,' says Charlton, who has been working in disaster response since 1988 and helped people after the Columbine massacre, as well as evacuees from Hurricane Katrina. 'Our job in disaster response is to help them find their resilience.'"[308]

Does it not remind you of the strategy of interpreters of the Bible or other "holy books"? When facts contradict claims written in such a book, it is their role to explain to followers that this specific claim is only a metaphor. New or other facts change other claims into metaphors, but the remaining content still holds "true." An endless story. And only interpreters can tell metaphors from "true" facts…

Bad things do happen to people. Some of them are unable to cope in many situations, so, understandably, they need help. Making thousands of people believe that life crises lead to PTSD and therefore they should seek psychological help has been a grave disaster and a cruel triumph of a cynical industry that shows no signs of ceasing. In the face of human misery, its representatives have no qualms in promoting the following perspective: "For survivors, psychological problems are more damaging than the earthquake."[309] And they will not stop hurting people.

An inquiry like the one conducted in this chapter throws me irreversibly into the abyss of despair. Am I really right? Have I read the relevant literature? Have I done my best? Was I unbiased when working out the arguments? The answer is a complex one, but if it is "no," that is: I am not right, I have read irrelevant material, I have not done my best, and I was in fact biased when pulling together the threads of my argumentation, then ... then there is something terribly wrong with my discipline. I have found two contradicting visions of the same concept, both created by representatives of my profession, and they simply cannot be reconciled. If there is no truth in my argument, it is even more urgent to hold fast to the belief that "we really ought to look into theories that don't work, and science that isn't science."

---

[269] N. deGrasse Tyson, "An Eye-Witness Account of the World Trade Center Attacks," *Prison Planet* (September, 2001): http://www.prisonplanet.com/eye_witness_account_from_new_york.html

[270] M. Ellison, Ed.Vulliamy, and J. Martinson, "We Got Down to the Outside and it was Like an Apocalypse," *The Guardian*, (September 2001): http://www.theguardian.com/world/2001/sep/12/september11.usa20

[271] R. Gist, and S. J. Woodall, "Social science versus social movements: The Origins and Natural History of Debriefing," *The Australasian Journal of Disaster and Trauma Studies 1*, (1998): http://www.massey.ac.nz/~trauma/issues/1998-1/gist1.htm

[272] J. T. Mitchell, "Protecting your People from Critical Incident Stress." *Fire Chief 36*, (1992): 61-67.

[273] A. C. McFarlane, "The Longitudinal Course of Posttraumatic Morbidity: The Range of Outcomes and Their Predictors," *Journal of Nervous and Mental Disease 176*, (1988): 30-39.

[274] K. Hytten, and A. Hasle, "Firefighters: A Study of Stress and Coping," *Acta Psychiatrica Scandinavia 355*, (1989): 50-55.

[275] D. A. Alexander, and A. Wells, "Reactions of Police Officers to Body Handling After a Major Disaster: A Before and After Comparison," *British Journal of Psychiatry 159*, (1991): 547-555; J. Thompson, and M. Solomon, "Body Recovery Teams at Disasters: Trauma or Challenge," *Anxiety Research 4*, (1991): 235-244.

[276] J. Griffiths, and R. Watts, *The Kempsey and Grafton bus crashes: The aftermath* (East Linsmore, Australia: Instructional Design Solutions, 1992).

[277] M. P. Deahl, A. B. Gillham, J. Thomas, M. M. Dearle, and M. Strinivasan, "Psychological Sequelae Following the Gulf War: Factors Associated with Subsequent Morbidity and the Effectiveness of Psychological Debriefing," *British Journal of Psychiatry 165*, (1994): 60-65.

[278] J. I. Bisson, and M. P. Deahl, "Psychological Debriefing and Prevention of Post-Traumatic Stress: More Research is Needed," *British Journal of Psychiatry 165*, (1994): 717-720.

[279] D. Meichenbaum, *A Clinical Handbook/Practical Therapist Manual for Assessing and Treating Adults with Posttraumatic Stress Disorder* (Waterloo, Ontario, Canada: Institute Press, 1994).

[280] R. Gist, and S. J. Woodall, "Occupational Stress in Contemporary Fire Service," *Occupational Medicine: State of the Art Reviews 10*, (1995): 763-787.

[281] B. Raphael, L. Meldrum, and A. C. McFarlane, "Does Debriefing After Psychological Trauma Work? Time for Randomised Controlled Trials," *British Journal of Psychiatry 310*, (1995): 479-1480.

[282] M. Hobbs, R. Mayou, B. Harrison, and P. Worlock, "A Randomised Controlled Trial of Psychological Debriefing for Victims of Road Traffic Accidents," *British Medical Journal 313*, (1996): 1438-1439.

[283] R. A. Mayou, A. Ehlers, and M. Hobbs, "A Three-Year Follow-up of a Randomised Controlled Trial of Psychological Debriefing for Road Traffic Accident Victims," *British Journal of Psychiatry 176*, (2000): 589-593.

[284] J. A. Kenardy, R. A. Webster, T. J. Lewin, V. J. Carr, P. L. Hazell, and G. L. Carter, "Stress Debriefing and Patterns of Recovery Following a Natural Disaster," *Journal of Traumatic Stress 9*, (1996): 37-49.

[285] I. J. Bisson, P. L. Jenkins, J. Alexander, and C. Bannister, "A Randomised Controlled Trial of Psychological Debriefing for Victims of Acute Harm," *British Journal of Psychiatry 171*, (1997): 78-81; R. Gist, "Is CISD Built on a Foundation of Sand?" *Fire Chief 40*, (1996): 38-42; R. Gist, "Dr. Gist Responds (Letter to the Editor)," *Fire Chief 40*, (1996): 19-24; R. Gist, J. M. Lohr, J. A. Kenardy, L. Bergmann, L. Meldrum, B. G, Redburn, D. Paton, J. I. Bisson, S. J. Woodall, and G. M. Rosen, "Researchers Speak on CISM," *Journal of Emergency Medical Services 22*, (1997): 27-28; R. Gist, B. Lubin, and B. G. Redburn, "Psychosocial, Ecological, and Community Perspectives on Disaster Response," *Journal of Personal and Interpersonal Loss 3*, (1998): 25-51; J. A. Kenardy, and V. Carr, Imbalance in the Debriefing Debate: What we Don't Know far Outweighs what we do," *Bulletin of the Australian Psychological Society 18*, (1996): 4-6; C. Lee, P. Slade, and V. Lygo, "The Influence of Psychological Debriefing on Emotional Adaptation in Women Following Early Miscarriage: A Preliminary Study,"*British Journal of Medical Psychology 69*, (1996): 47-58; S. Rose C. Brewin B. Andrews and M. Kirk "A Randomised Controlled Trial of Individual Psychological Debriefing for Victims of Violent Crime," *Psychological Medicine 29*, (1999): 793-799; C. Stephens, "Debriefing, Social Support, and PTSD in the New Zealand Police: Testing a Multidimensional Model of Organizational Traumatic Stress," *Australasian Journal of Disaster and Trauma Studies 1*, (1997): http://massey.ac.nz/~trauma/issues/1997-1/cvs.htm; J. Kenardy "Current State of Psychological Debriefing," *British Medical Journal 321*, (2000): 1032-1033.

[286] R. Gist, B. Lubin, and B. G. Redburn, "Psychosocial, Ecological, and Community Perspectivesr Response," *Journal of Personal and Interpersonal Loss 2*, (1998): 25-51.

[287] S. Rose, and J. Bisson, "Brief Psychological Interventions Following Trauma: A Systematic Review of the Literature," *Journal of Traumatic Stress 11*, (1998): 697-710.

[288] S. Rose, Evidence-Based Practice will Affect the Way we Work," *Counselling 3*, (2000): 105-107.

[289] S. Wesley, S. Rose, and J. Bisson, "A Systematic Review of Brief Psychological Intervention (Debriefing) for the Treatment of Immediate Trauma Related Symptoms and the Presentation of Post-Traumatic Stress Disorder," *(Cochran Review)*, (Cochrane Library: Issue 3, Oxford, UK: Update software: 1998).

[290] S. Wesley, S. Rose, and J. Bisson, "Brief Psychological Interventions ("Debriefing") for Trauma-Related Symptoms and the Prevention of Posttraumatic Stress Disorder," *(Cochrane review)*, The Cochrane Library, 4 (on-line), (2000): http://www.cochranelibrary.com/cochrane/cochrane-frame.html

[291] For a review, see: J. M. Lohr, W. Hooke, R. Gist, and D. F. Tolin, "Novel and Controversial Treatments for Trauma-Related Stress Disorders," in *Science and Pseudoscience in Clinical Psychology*, eds., S. O. Lilienfeld, S. J. Lynn, and J. M. Lohr, (New York: Guilford, 2003): 243–272.
[292] B. Carey, "Sept. 11 Revealed Psychology's Limits, Review Finds," *The New York Times*, (July 28, 2011):
http://www.nytimes.com/2011/07/29/health/research/29psych.html?ref=todayspaperand_r=0
[293] Ibid.
[294] J. Stevenson, "Therapy can Drive you Mad, Finds Study on Counselling Given to 9/11 Survivors," *Mail Online*, (July 31, 2011):
http://www.dailymail.co.uk/news/article-2020699/Therapy-drive-mad-finds-study-counselling-9-11.html
[295] S. K. Whitbourne, "How to Manage Traumatic Reactions to Disasters," *Psychology Today*, (August 30, 2011): https://www.psychologytoday.com/blog/fulfillment-any-age/201108/how-manage-traumatic-reactions-disasters
[296] Ibid.
[297] V. Bell, "Minds Traumatised by Disaster Heal Themselves Without Therapy," *The Guardian*, (May 12, 2013):
http://www.theguardian.com/science/2013/may/12/natural-disasters-healing-psychology-worse
[298] D. Hawker, J. Durkin, and D. Hawker, "To Debrief or not to Debrief our Heroes: That is the Question," *Clinical Psychology and Psychotherapy 18*, (2011): 453-463.
[299] Ibid.
[300] C. Jarret, "Is it Time to Resurrect Post-Ttrauma Psychological Debriefing for Emergency Responders and Aid Workers?" *Research Digest*, (January 16, 2012): http://digest.bps.org.uk/2012/01/is-it-time-to-resurrect-post-trauma.html
[301] E.g.: http://www.cima.org.au/education/educators
[302] P. J. Watson, M. J. Brymer, and G. A. Bonanno, "Postdisaster Psychological Intervention Since 9/11," *American Psychologist 66*, (2011): 482–494.
[303] Ibid., 485.
[304] K. E. Paul, and M. van Ommeren, "A Primer on Single Session Therapy and its Potential Application in Humanitarian Situations," *Intervention 11*, (2013): 8-23.
[305] J. McGill, "How Critical Incident Stress Debriefing Teams Help Cops in Crisis," *PoliceOne.Com*, (May 22, 2015): http://www.policeone.com/health-fitness/articles/8554063-How-critical-incident-stress-debriefing-teams-help-cops-in-crisis/
[306] "Critical Incident Stress Debriefing and Mythology," *Rouge Medic*, (November 10, 2010): http://roguemedic.com/2009/11/critical-incident-stress-debriefing-and-mythology/
[307] World Health Organization, "Psychological Debriefing in People Exposed to a Recent Traumatic Event,"*Mental Health*:
http://www.who.int/mental_health/mhgap/evidence/other_disorders/q5/en/
[308] J. Levs, "Grief Counselors Offer Ssolace Amid Tragedy," *CNN*, (July 24, 2012): http://edition.cnn.com/2012/07/23/us/colorado-grief-counselors

[309] C. Sharma, "For Survivors, Psychological Problems More Damaging than the Earthquake," *AsiaNews.It,* (May 20, 2015): http://www.asianews.it/news-en/For-survivors,-psychological-problems-more-damaging-than-the-earthquake-34295.html

Chapter 10: Experimental Therapy Patient's Handbook

Despite the numerous warnings I have included in my books, many people still want to embark on the adventure of psychotherapy. Many potential candidates for patients of therapies with effectiveness proven by EBP – as well as those still in the phase of uncontrolled experiments on human subjects – probably wonder how they should go about choosing the most appropriate therapy and therapist for them. I will try to assist them and, in as simple and understandable a manner as possible, provide guidelines and selection criteria that may be helpful both in choosing the therapy and a particular therapist.

The first, absolutely fundamental notion that should be verified before undergoing any type of psychotherapy is making sure if research results – naturally, carried out in accordance with APA guidelines, about which I have already written in previous sections of this book – confirm the effectiveness of the given method. If the therapist is not able or does not want to answer this question, find another one. They either do not know the answer – which means it is probable they are yet another cargo cult follower – or they know the answers are less than impressive. By not working with such therapists you avoid many possible problems.

Apart from the primary consideration of examining evidence of effectiveness, there is also the notion of *being aware of the possibility of asserting one's rights in problematic situations*. Before we knock on the door of the therapist's office, regardless of how esteemed they might be, it is vital to ask what possibilities you have of receiving compensation in the event of any harm meeting you during therapy. When buying a car, a refrigerator, a bike, or even shoes, we want to be sure of the rules regarding service or warranty. We do not ask for those at the moment of purchase because our intent is to return the bought goods, but rather because we want to feel safe. The conditions of warranty and reimbursement in retail, services, or even medicine, regardless of how they are exercised, are still quite precisely regulated by law. In the case of psychotherapeutic services, there are not always legal means of asserting claims. I personally know a few examples of prosecutors who, despite gross malpractice by therapists, have discontinued inquiries due to the impossibility of proving guilt.

However, this scenario is not only a likely one in my country, Poland. As chair of the UK Council for Psychotherapy, Professor Andrew Samuels, once said to a journalist: "Imagine a friend of yours is in therapy with me and I make a pass at her. At the moment she can complain in confidence and we believe she will get a fair hearing [in private]. But in the future she would face enormous difficulties. Her complaint would be trampled over by committees, and she runs the risk that everything I know about her will end up in the newspapers. We feel that, in public protection terms, it's a crappy system."[310]

This is the bad news. The good news is that among therapists there are honest people with good intentions, doing their best to maintain the prestige of the psychotherapy profession. These people try to create substitutes for legal rules that are meant to serve the patient. For that reason, several types of Patient's Ombudsman have been created, which, in medicine (in retail and services there is a consumer rights ombudsman), provide help in critical situations and represent patients/clients. A good example of such an ombudsman is the Parliamentary and Health Service Ombudsman in England.[311] In the United States, each state has its own set of laws that regulate the practice of health care professionals. Traditions regarding this issue vary immensely in different countries, and therefore it is difficult to identify a single path of action for a mistreated patient.[312] When beginning therapy, it is important to ask whether there is a possibility of seeking the help of an ombudsman. If so, who? If there is not, then why? What is the opinion of the psychotherapist we have chosen on this subject?

A type of substitute for such an office, and at the same time the basis of asserting one's rights, can be the code of ethics or code of good practices of the organization our therapist belongs to. Therefore, it is important to check if they are a member of any professional society, and to read any literature detailing such codes. It is also good to understand the procedures for filing and reviewing complaints regarding unethical therapist behavior, and how they are implemented. Colin Walker, policy and campaigns manager at the mental health charity Mind in the UK, warns: "There are a plethora of organizations with their own procedures, and no minimum trading standards for the industry. Some people have waited years for their complaints to be processed, and the end sanctions can be minimal. What's more, those registers are voluntary, and you can continue to practice."[313]

A complementary element to official procedures can be establishing and maintaining a website where patients may express their opinions regarding psychotherapists. In fact, the more transparency in the operation of a given school of therapy and openness to feedback, the better. Unfortunately, my subjective perception indicates that the further a given school is from EBP, and the more eagerly it calls itself "art," the fewer patient protection mechanisms it provides. Well, in murky waters…

Another important warning – pseudotherapies present on the market long enough are not only able to imitate science, but also simulate the presence of the discussed protection tools. When deciding to undergo therapy, one should remember that we are choosing from at least a few hundred schools and modalities, many of which are in conflict with one another. Only a handful of them can boast empirically proven effectiveness. However, this is not all that should be examined.

A very important element in the work of therapists is the *system of supervision* employed in a given school of therapy. In order to explain what this is, I will tell you about my first encounter with this procedure. It happened long ago, in 1993. I was on a scholarship program at Bielefeld University, Germany. A colleague I worked with in the same room proved to be not only a scientist, but also a practicing psychotherapist. One weekend, he was going to nearby Osnabrück, for a convention of therapists from his school. He asked if I wanted to go with him. I agreed, on the one hand wanting to take a closer look at the work of German psychotherapists, and on the other I was lured by the opportunity to visit an old city full of historic buildings.

The meeting of eight therapists took place in a large, private apartment. It began with all the participants going to their bags to retrieve containers the size of shoeboxes, filled with audio cassettes. These cassettes contained records of all the therapy sessions carried out by everyone present since their last meeting. I would like to stress the fact that all of the therapy sessions were recorded! After agreeing on the order in which they would proceed, the group review was begun. The first stage was to randomly draw a cassette from the box, listen to it, and conduct a feedback session for the therapist in question. In the second stage, the therapist asked the group to listen to the fragments he was unsure of, or which he felt were problematic. This stage also ended in discussion and feedback. The meeting lasted over a dozen hours, and such sessions took place at least once a month.[314]

Apart from the broad array of benefits the therapist could gain from such meetings, it is vital to stress the evident benefits for the patient. First of all, they gain a certain guarantee that any errors on the part of the therapist can be detected. The supervision system also means that the therapists of a given modality continually strive to develop their skills and expertise. Moreover, what is really important, transparency of the therapeutic process eliminates any purposeful attempts at breaking ethical rules by the therapist, and such cases, as we described in *Psychology Gone Wrong*, are quite frequent.

Although numerous authors indicate the difficulties and limitations arising from supervision,[315] these are problems that affect the therapists themselves, and it would be quite hard to show any negative effects that implementing such procedures could have on the patient.[316] Therefore, the next questions one should ask before undergoing therapy should relate to whether it is supervised, how often, and in what form. While the mere fact of supervision does not unambiguously determine the value of a given therapy, it is highly probable that its absence is a sign that we are dealing with charlatanism, or at least with practices containing a high dose of subjective arbitrariness in the interpretation of proper procedure.

In the Woody Allen movie *Sleeper*, the main protagonist, having learned that he has been asleep for two hundred years, is predominantly troubled by the fact that he has lost so much time – if he had been able to continue to attend therapy sessions through those years, he surely would have come close to finishing by now. Whether Woody Allen was exaggerating or not, it is a fact that the process of therapy is rarely short (with the exception of short-term therapy). One should be prepared for a program that lasts months and years, rather than days and weeks. And since time is not controlled by anyone – indeed, it flows according to its own rhythm and brings unforeseen events – before beginning therapy, it is important to consider two types of situations that may likely arise and ask the therapist about them. Firstly, people sometimes experience financial difficulties. What will happen if, during the therapy, I lose my ability to pay for it? Will the therapist continue the sessions, or will he terminate them in accordance with the laws of the market? Ask how either of the situations could affect both you personally as the patient and the course of therapy itself.

Fortunately, as well as hard times, there are also positive events in life such as achieving a promotion, getting a dream job, receiving an inheritance, or winning the lottery. Let us consider a less random event – you got a promotion and your company is sending you to another city or country in order to manage its new office there. You are in the middle of therapy. Does your therapist record the course of therapy in a manner enabling him to transfer you as a patient to another specialist? Does he follow precise procedures and standards of operation so that the new therapist will be able to determine what has been done and what still needs to be addressed? Did he define the purpose of the therapy, and is he able to determine when the goals will be achieved? Or, rather, does he treat the whole process as a creative act, in which you are the medium, and nobody apart from him is able to "complete the masterpiece"?

Ask your therapist these questions. His answer can paint a picture for you of what kind of procedures you will face, and help you make the final decision. However, this is still not the end of the verification process.

Another element worth checking before beginning psychotherapy is the therapist's insurance. Liability insurance in the case of a therapist concerns any damage resulting through the broadly-understood healing process. There does not have to be any permanent damage to the patient's health, let alone death, since the patient can also demand compensation in the case of a needlessly prolonged therapeutic process, or in the event of suffering side effects, as well as in the case of not meeting established deadlines in the therapeutic service, erroneously completed documentation, and many other situations.

Liability insurance in conjunction with operating a business that provides non-material services is standard protection against inadvertent injury. This is how accounting offices, advisory companies, doctors' offices, and so on are insured. An error in therapy or mistakes in organization may happen to any of us. Nothing stands in the way of a psychotherapist insuring himself against such events, especially since patients' legal awareness is continually on the rise, and they are increasingly conscious of the fact that in the event of malpractice they are entitled to pursue their claim in court. An insured therapist gives the patient more certainty that they will be able to successfully claim compensation for any damage sustained during therapy.

Although both in the US and the countries of Western Europe professional insurance is a widely-used tool, this does not always amount to patient security. Such was the situation Dawn Devereux, previously a harassed patient, now also a therapist, found herself in.

> Dawn Devereux was slightly nervous when she started seeing a psychotherapist, but she never anticipated the experience would descend into sexual come-ons by someone she knew to be a respected professional. "It left me traumatised," she says of her six years with the therapist. "Very quickly it became strange. He talked about himself all the time. He didn't listen to me. He got angry when I questioned him. After a while, he would say I needed a cuddle. Then, a few years into it, he said he wanted to have sex with me."
> Devereux says she finally found the courage to walk away from the therapy, and complained to the professional body to which the therapist belonged – only to be greeted with a tribunal system "run by his colleagues" and to be told that if the panel found in favour of the therapist she would have to pay all the costs.
> "I just couldn't believe it," she recalls. "It's difficult to explain how I felt [while in therapy], but I was vulnerable, I suppose. He was behaving inappropriately and I didn't know what to do. I put myself through the complaints process of his professional body, and then was told I might have to pay their costs if I lost."[317]

In Poland and other countries of Central and Eastern Europe most therapists do not trouble themselves with any kind of insurance. There may be various reasons for this. Firstly, ignorance – they may not be aware of such a possibility and necessity. Secondly, unawareness or underestimating the risk – they may have never heard of a psychotherapist's patient seeking to assert her rights in court.

Another possible reason is the deliberate exploitation of the upper hand that a therapist has in such situations. It may be the case that some representatives of this profession are aware of the possibility of insuring oneself, as well of the risk of committing an error and (unintentionally) causing harm, but still decide not to take the opportunity to buy a policy, as they are almost certain that patients will not able to prove any error or omission on their part. Another reason why therapists do not sign insurance agreements - one quite dangerous for patients - arises from the opinion that even if psychotherapy does not help, it also cannot bring significant harm to anyone. And the worst possible motive of a psychotherapist who does not even consider liability insurance is the certainty of being infallible. Unfortunately, such cases really do happen – I personally know psychotherapists convinced that they do not make mistakes.

Nonetheless, when deciding to put your fate in the hands of a chosen psychotherapist, it is still worth asking about insurance, or for the therapist's reasons for not having any. It is true that even if we meet an exceptionally cynical therapist, we might not hear an admission that there is no sense in insuring oneself in a situation when guilt cannot be proven, but the way the answer is given may provide us with information.

Discussion of the issue of insurance should also be complemented with data we have mentioned in *Psychology Gone Wrong*, information that comes from people who have their feet on the ground, a group which most certainly includes representatives of insurance companies. For quite some time now, insurance providers in the USA have refused to insure psychotherapists against claims of sexual abuse, and this is not only due to stereotypes or prejudice. A psychotherapist is a client similar to any other, but in this case the decision was determined by numbers. Accusations of sexual harassment are the leading type of negligence claims in terms of the value successfully pursued in court proceedings against American therapists in the period 1976–1986. As much as 44.8% of the total sum of adjudged damages was paid for this particular claim. Therefore it is not surprising that psychotherapists are the only US professional group which insurance companies refuse to conclude insurance agreements with owing to the frequency of claims filed for sexual harassment during professional performance.[318]

If you have succeeded in finding a therapist who meets all of the above conditions, you should ask him one more question regarding his belief system. Why? Regardless of what the therapists themselves might tell you, numerous research has shown that during the process of psychotherapy – even if carried out using the least directive methods – modifications, or even complete changes to the patient's system of values take place, and such modifications are always in the direction of greater compatibility with the belief system espoused by the therapist.[319] Therefore, if you are a Catholic who cherishes the Catholic system of values and you want to take part in family therapy, choose a therapist who holds similar values. Otherwise, after a year of therapy it may turn out that you have accepted solutions you would not have accepted before therapy for ideological reasons. Likewise, if you are an atheist, avoid "therapists of the cloth," and if you are a feminist, find a therapist who values this worldview.

Summary – a list of questions to ask before deciding to begin therapy:

1. Have the methods used by the therapist been the subject of any type of research in accordance with APA guidelines?
    a) YES: Did the research confirm the effectiveness of any of these methods?
    b) NO: Why not?
2. What are the options for asserting rights in problematic situations that may arise during therapy?
    a) Is there an option for the patient to report problems to any kind of ombudsman?

b) Is the therapist a member of a professional association which has its own ethical code, and does he respect it?
   c) What are the formal options for demanding compensation in the event of violating the code (association's ethics committee, peer arbitration, etc.)?
   d) Are there any known cases of complaints and evidence of the manner in which they were addressed?
   e) Is there an Internet forum or any other platform where patients can share opinions regarding therapists and the therapeutic process? Does the therapist accept and support such a form of sharing information?
3. Does the school of therapy that the therapist follows have a supervision system?
   a) YES: What does the system look like? How often is the work of the therapist supervised?
   b) NO: Why?
4. What will happen if, during your therapy, you lose your ability to pay for it?
   a) Will the therapist continue the meetings, or will he discontinue them, in accordance with free market rules?
   b) What impact may such a situation have on you as a patient and on the course of therapy?
5. Will your therapist keep records of the course of therapy in a manner that would enable him to transfer you as a patient to another therapist in the event of any unexpected events?
   a) Will therapy be carried out based on precise procedures and operating standards in a manner that will enable control over what has already been done and what still needs to be addressed?
   b) Will he formulate, together with you, the objective of the therapy and be able to determine the stage which it has reached?
   c) Or, rather, will he treat the whole event as a creative act in which you are only the artistic medium and nobody but he can "finish the masterpiece"?
6. Does the psychotherapist carry liability insurance?
   a) YES: What situations does the insurance agreement cover?
   b) NO: Why not?
7. What belief system does the therapist espouse?
   a) To what extent is it compatible with your own?

If the answers to all of these questions are in accordance with the suggestions I have discussed in this chapter, and they do not raise doubts, you can assume that you have done your very best in order to protect yourself from a harmful course of therapy. Now all that remains is for me to wish you luck!

[310] J. Doward, "Sex Scandals, Rows and Mavericks: Is it Time to Regulate Psychotherapy?" *The Guardian,* (May 9, 2010):
http://www.theguardian.com/lifeandstyle/2010/may/09/rogue-psychotherapy-regulation-row
[311] http://www.ombudsman.org.uk/
[312] For more see: P. Jenkins, ed., *Legal Issues in Counseling & Psychotherapy* (London: SAGE Publications, 2002).
[313] Doward, "Sex Scandals."
[314] Naturally, supervision doesn't always have to take such a form. In many schools of therapy supervision boils down to work with only one supervisor, and is closer to the formula first proposed in 1920 by a psychoanalyst from Berlin, Max Eitington (comp. B. Hutto, "Some Lessons Best Learned from Psychotherapy Supervision," *Psychiatric Times 18,* (2001): 7). The model described above should rather be called a Balint group with a supervisory function (M. Balint, *The Doctor, his Patient and the Illness* (London: Pitman, 1964).
[315] C. Bailey, "Do we need Supervision?" *PsychotherapyTodey.net,*
http://www.therapytoday.net/article/show/3475/do-we-need-supervision/
[316] The only situation that comes to mind is one in which the student has long outgrown the master (supervisor), who is being lead onto a wrong path. Although such a situation is possible in a teacher-student relation, it is far more probable that the teacher possesses more knowledge and experience then the student.
[317] M. O'Hara, "Thought policing," *The Guardian,* (July 1, 2009):
http://www.theguardian.com/society/2009/jul/01/dawn-devereux-psychotherapy
[318] R. Z. Folman, "Therapist-Patient Sex: Attraction and Boundary Problems," *Psychotherapy 28,* (1991): 168-173.
[319] A. Bandura, D. H. Lipsher, and P. E. Miller, "Psychotherapists' Approach-Avoidance Reactions to Patient's Expressions of Hostility," *Journal of Counsulting Psychology 24,* (1960): 1-8; E. J. Murray, "A Content Analysis Method of Studying Psychotherapy," *Psychological Monographs 70,* (1956): 420; M. B. Parloff, "Some Factors Affecting the Quality of Therapeutic Relationships," *Journal of Abnormal Social Psychology 52,* (1956): 5-10; R. Rosenthal, "Changes in Some Moral Values Following Psychotherapy," *Journal of Counseling Psychology 19,* (1956): 431-434.

PART III:
IS THERE ANYTHING YOU COULD NOT DO FOR YOUR CHILD?

Preying on which human misfortune could make the most money? Confronted with this truly awkward and disturbing question, many of us might start our lists with human trafficking, organ trading, or peddling useless drugs to the unwitting. Given time to think, though, we would likely rearrange that list a bit, topping it with those who exploit children as soldiers, sex industry workers, or forced labor, but above all with those who prey on their mental conditions or disabilities. While doing this mental exercise we would have tended to place images associated with these horrible things up in our heads as taking place somewhere far away: in the developing world, in Asia or Africa, places where one does not visit on a package tour.

Meanwhile, right at the heart of our smug and carefree Western world, an industry has been flourishing that targets just that kind of misery experienced by children and their parents: disabilities, mental conditions, and sicknesses of the nervous system. That we fail to notice it on a daily basis is testimony of the clever ways in which it operates. Indeed, it has not only avoided attracting universal outrage, but also managed to enjoy the occasional endorsement from such established and respected organizations as the European Union, UNESCO, government agencies in many countries, and even some universities.

It might seem obvious that universal efforts are needed to battle a handful of major threats faced by humankind, such as cancer, AIDS, and ailments that affect children, who are by nature innocent and powerless. The reality, however, resembles more an image from nature documentaries: when an animal becomes weak and vulnerable, incapable of defending itself and/or dying, it is immediately surrounded by a pack of hyenas or a flock of vultures. Before the victim draws its last breath, the carrion eaters already begin fighting for position to secure access to the juiciest pieces.

This is precisely the situation that befalls an incurably ill person in our supposedly enlightened civilization. As soon as the horrible truth descends, as soon as the patient knows what awaits, miracle workers and representatives of assorted alternative medicines and therapies start flocking in with promises flowing from their lips, deep concern and sympathy on their faces, and the venom of false hope ready to be administered into the victim's weakened mind. Their utter lack of scruples makes it all the harder for others who are in a position to offer genuine help or at least to ease the suffering. Indeed, a proper doctor will not dare promise the impossible, and an honest psychotherapist will be very careful in formulating his prognosis, even if he realizes the healing potential of hope. Yet alternative miracle workers, to whom this chapter is devoted, have no such qualms and will promise anything. How, then, can an honest therapist survive among them? Who will buy the modest promises he makes and the restrained hope he offers?

In the wild, hyenas do not camouflage their intentions by carrying tasty morsels in their mouths while approaching their prey. The human hyenas, on the other hand, approach their victims with the sweetest talk. To them the longer the victim takes to suffer before it dies the better, because this increases the opportunity to profit. Mentally handicapped and mentally ill children are particularly valued, because they will live long while their healthy parents work themselves to death to keep their hopes alive. Naturally, the hopes do not come free and yet the miracle workers risk nothing. Think about it: would you blame someone who has devoted his or her entire life to children? Even an onlooker unaffected by misfortune and perhaps lacking the basic understanding of the method used by such child therapists will rather tend to respect, recognize, and reward their devotion than attack them.

Let us now have a closer look at a few such candidates for sainthood.

Chapter 11: A Saint or a Charlatan?
The Doman-Delacato Method

What sacrifices are parents ready to make once they have learned of their newborn child's mental disability, or parents who fear that their progeny may suffer from some kind of impairment or mental illness? What kind of emotional toll does it take on parents when their initial suspicion and fear turns into certainty? How often does the feeling of guilt plague them? What comforts will these unfortunate people renounce for the sake of their children? What will they sacrifice once they become persuaded that the impossible is in fact possible, that is to say, everything can be cured and that there are scientifically validated methods to overcome the disabilities their children are faced with? The self-evident answers to these questions must have been a source of inspiration to Glenn Doman at the time when he founded The Institutes for the Achievement of Human Potential (IAHP) based in Philadelphia and announced that: "Our individual genetic potential is that of Leonardo, Shakespeare, Mozart, Michelangelo, Edison, and Einstein. ... Our individual potential race is not that of our parents or grandparents. ... All intelligence is a product of the environment."[320]

And although the claims quoted above – similarly to many of Doman's contentions – stand in apparent contradiction to scientific research findings, they may be a source of fervent hope for any individual struggling with a personal tragedy.

Glenn Doman, as befits a paragon of a positive hope, had the appearance of an avuncular wise man, with his gray hair and matching gray beard. He often had his picture taken surrounded by happy children, or, occasionally, in the company of parents whose eyes shone with boundless adoration and gratitude. Less frequently, the pictures show him surrounded by his devoted followers or at a lectern. There are no hints of bad character in his face - it exudes kindness and readiness to bring help to others. He publicly opposed his therapy being advertised as the "Doman method" because, as he explained, dozens of people had contributed to the formation of the theory and treatment methods. He built up an image of an unselfish and modest person. When relating his success story in books, Doman seemed almost surprised by his achievements: "While at first this seemed an impossible or at least monumental task, in the years that followed we and others found both surgical and nonsurgical methods of treating the brain itself."[321]

What were the underlying motives for the development of his pseudoscientific machine? Had Doman fallen victim to his own misjudgments when formulating an inconsistent concept that consequently raised as much hope as disillusionment? Or perhaps he was driven by cynical calculation, expecting to achieve wealth and fame in return? In order to find the answers to these questions, let us take a look at some episodes from his past.

We begin in 1940. In this year, Glenn Doman completes his studies in physical therapy at the University of Pennsylvania. He develops his methods in the mid-1950s. An early influence was Temple Fay, a famous Philadelphian neurosurgeon with extensive expertise in rehabilitation treatment for children suffering from cerebral palsy. Glenn Doman, as a member of Fay's professional team, takes his mentor's concepts to heart, and, together with educational psychologist Carl Delacato, develops them to formulate his own method of treating brain damage in children. It is hoped that this treatment will cure almost all permanent psycho-degenerative conditions in even the youngest patients, including cerebral palsy, intellectual disability, autism, and Down syndrome. With a view to putting the method into practice, he later establishes a children's physical rehabilitation center in Philadelphia. Before long, an entire network of such centers comes into being under the proud name of The Institutes for the Achievement of Human Potential (IAHP), As a rule, the Institutes operate on a *non-profit* basis. In 1960, the first and, until today, only scientific article providing an overview of the method is published.[322] In the subsequent years Doman makes "discoveries" that, as he claims, enable him to work with non-impaired children too in order to increase their intellectual potential. In 1964, Doman's first book is brought out by a publishing house established by the IAHP and bearing an extremely telling name, The Gentle Revolution Press. Doman then travels around the world and works intensively with children of more than a hundred nationalities. The year 1966 sees him being honored with one of Brazil's highest decorations. In 1974, Doman's next book comes out, *What to Do About Your Brain-Injured Child*,[323] probably the most fundamental text in the context of his approach. In 2007, at the invitation of former President of the Soviet Union Mikhail Gorbachev, Glenn Doman attends the Pio Manzù International Conference held in Rimini, Italy, and is awarded the Medal of the Italian Senate in recognition of his dedicated work in the field of child brain development.[324]

If we now jump forward to the present, we can see that the Doman-Delacato method, as it is most commonly called, is being applied all around the world. Outside the United States, IAHP offices also operate in Italy, Japan, Mexico, Guatemala, Singapore, Brazil, Spain, and France. In addition, in places where official branches have not been established, advocates of the method have successfully overseen its inclusion in a number of institutions offering physical rehabilitation, education, and therapy services.

It is worthwhile reviewing the concept that not only has brought world recognition and fame to its author, but, in consequence, has become a source of immense profits for him. It is rooted directly in the recapitulation theory, upheld by Temple Fay, which, in short, proposes that in the course of ontogenesis (growth from an embryo to adulthood) specific stages of the phylogenetic development of a given species recur. Based on this assumption, Fay had theorized that the human brain must analogously undergo development stages that correspond respectively to the brains of fish, amphibians, reptiles, and other mammals, before it eventually turns into a typically human form of brain. Doman concluded that if this process really takes place, a "brain injury" experienced at a particular stage of the neurological development prevents the brain from growing properly. Followers of the Doman-Delacato method draw on the theory that the brain is innately capable of developing both functionally and anatomically. Furthermore, IAHP therapists maintain that a conventional treatment based on drug administration causes brain damage and brings about negative side effects.[325] At the same time, they maintain that owing to neuroplasticity, their programs of sensory stimulation may in fact trigger the physical development of the brain and lead to "the emergence of improved neurological functions" in patients.

Another important thesis in Doman's concept is the claim that a shortage of oxygen in the brain is the key underlying cause of many problems experienced by children who suffer from brain damage. Therapists of the Institutes therefore proudly assert that their programs encompass techniques that help provide the brain with a sufficient supply of oxygen, thus facilitating the health recovery process.

At this point it already seems necessary to explain the origins of the term *injury*, a key part of the theory here. The truth is that Doman deliberately uses self-made phrases, such as "neurological injury," "neurological dysorganization," or "dysfunction" (this is how he defines even the lack of one cerebral hemisphere!) to devise his own categories. He does so with the aim of avoiding having to employ the formal diagnostic terminology and such phrases as handicapped child, special child, spastic, or autistic. Typical of this technique, the use of the term *injury* is supposed to imply the promise of *healing*.

Such tenets have led to the formation of a therapeutic scheme that incorporates a variety of measures in the following areas:
- physical development
- physiology – factors that affect physical development, such as water and electrolyte management, blood gas saturation, etc.
- intellectual development
- emotional and social development.

The therapeutic procedure outlined by the Doman-Delacato method begins with a so-called functional analysis. It is intended to determine, based on symptoms, the degree of impairment or to confirm the lack of activities regulated by relevant functional brain levels. Next, the location, size, and severity of the "brain injury" is identified with use of a tool invented in the IAHP, i.e. the *Developmental Profile*. The subsequent stage takes the form of an entire therapeutic process that involves measuring six neurological "competences":
- visual
- mobility

- auditory
- language
- tactile
- manual.

One should be aware that my depiction of the Doman method is not exhaustive and is only meant to provide a representation of the therapeutic process. For this reason I will confine myself to delineating only some of the therapeutic activities and techniques, without claiming to present the method in its entirety.

One of the method's most vital rehabilitation components is *patterning*, which springs from the premises of the recapitulation theory. As I have explained above, the theory holds that evolutionary (phylogenetic) transformations also occur in an individual's development (ontogenesis). Therefore, the child's motor development is believed to progress through certain stages, with the lowest being "homologous crawling," where the child's limbs move forward at the same time (a frog-like movement pattern). The process then continues, next reaching the "homolateral crawling" stage, where the upper and lower limbs on the same side move forward at one time (a salamander-like movement). The final stage of mobility development is "cross-pattern creeping," that is, alternate movements of the upper limb of one side and the lower limb of the opposite side (a lizard-like movement). Patterning is simply a technique in which such movement patterns are passively reproduced. The patterning sessions themselves are conducted by a group of several people who manipulate the limbs of the child in a rhythmical manner, with the child being placed on the floor or on the table. The application of patterning, as Doman himself emphasizes, must be frequent, rigorous, and conducted on a regular basis. In his view, it is only through intensive training that the brain may be healed. This is how supporters of the method evaluate the activities themselves:

> Difficulties with the promotion and consistent use of this outstanding method stem primarily from the necessity to allocate substantial financial outlays and from the organization of work itself because rehabilitation of just one child requires the involvement of many people at the same time. To illustrate how laborious and time-consuming the method is: in a certain Swiss town, 240 people (volunteers) were engaged to do exercises with one child. The trainers, working in groups of five, took turns every two hours to do a 12-hour rehabilitation cycle. Their shifts were scheduled every eight days. Such a rehabilitation process goes on for many years.[326]

Another element of the therapy is so-called *masking*. In this technique, the child, with a small plastic mask placed on its face, is forced to inhale a mixture of air with an increased quantity of carbon dioxide. Masking takes from 30 to 120 seconds and is repeated between a dozen and several dozen times a day. This therapy is compliant with the proposition that many abnormalities in the functioning of the brain are caused by hypoxia. Air enriched with carbon dioxide is believed to improve blood circulation, reduce spasticity of muscles, and increase the control of involuntary movements.

An array of recommendations includes, among other things, hanging the child upside down or somersaulting, as well as executing *bit* programs designed to enhance the child's intelligence capacity with the use of so-called bits (*bits of intelligence*). This technique involves exposing the child to repetitive demonstrations of themed sets of cards with relevant images on them. Doman attests that by doing so, it is possible to raise the intelligence quotient from zero (the vegetative level) to 100 (average intelligence level), as well as to teach intellectually impaired two-year olds (!) reading and counting. The Doman therapy requires the administration of vitamin supplements as well as strict adherence to a dietary regime with prescribed daily amounts of water, glucose, and dairy products.

The Doman-Delacato method offers a promise that most disorders and damages of the nervous system may be cured, most of all cerebral palsy, Down syndrome, Asperger's syndrome, and autism. There is also a separate program, likewise stemming entirely from the underlying principles of the Doman-Delacato method, that is said to be effective in treating epilepsy. To my regret, I have not managed to track down either a list of contraindications or an explicit enumeration of disorders whose treatment based on the method can be ineffective.

As regards development opportunities for non-impaired children, the method lures parents with a promise of amazing growth in their intellectual potential and skills. Doman himself makes the following reference to this claim on the first page of one of his books: "Today there are tens of thousands of children ranging from babies to young adults who learned to read at early ages using this book."[327]

Let us have a closer look at this revolutionary approach. To this end, we need to reflect back on its theoretical tenets. The recapitulation theory, which constitutes the foundation of the Doman-Delacato method and acknowledges the reproduction of all evolutionary stages of particular species in the course of the individual development of organisms, was formed in the 19th century. Its central thesis was *the biogenetic law*, occasionally referred to as Haeckel's law, and was subsequently adopted in an oversimplified version by Doman and his followers. It can be summarized in the phrase "ontogeny recapitulates phylogeny." Although in biology and related sciences evolution had been customarily seen through the lens of biogenetic law for years, the first doubts surfaced as early as in the 19th century. Today, the recapitulation theory is mentioned almost entirely in its historical context, in the same way as the Lamarckism theory, Lysenko's concepts, and other ideas that have long been discarded by science. Moreover, even the most fervent supporters of the recapitulation theory never claimed that the individual's psychomotor development re-enacted the earlier evolutionary patterns. These facts, however, seem to have no effect on either the authors of the concept reviewed in this section or the therapists who put it into practice and that is why thousands of children worldwide, assisted by their parents and therapists, laboriously emulate the movements of frogs, salamanders, and lizards.

Could Temple Fay, Glenn Doman, and Carl Delacato have been unaware of evidence that questioned the truthfulness of the biogenetic law? There can only be one explanation for this – they all must have been utter ignoramuses with no knowledge of the scientific achievements of the entire last half-century. This, however, is contradicted by their biographies. From his graduation until 1943, Fay was an active researcher and served as the head of the Department of Neurosurgery, and Doman participated in his scientific investigation programs. Therefore, there is one hypothesis left, i.e. that they deliberately adopted erroneous assumptions while setting up their pseudoscientific concept. This perspective is also supported by the manner in which they chose to develop their method, i.e. outside the academic environment and the scientific press, and slamming the door in the face of interested scientists.

I am convinced that at this point enthusiasts and practitioners of the Doman-Delacato method, indignant over this argumentation, would accuse me of dishonesty in my criticism of the assumptions. After all, it is not theory that counts, but practice, in particular one that brings such amazing results! I am indeed far from acknowledging the primacy of practice over theory; however, I shall be glad to investigate practice as well, just as many scientists have done in the fields of physical rehabilitation, therapy, and treatment. Any method that is brimming with far-reaching promises, should, in the long run, initiate a true breakthrough in psychology, medicine, and education.

These promises are implied in the first sentences of Doman's books, even in the apparently not persuasive dedication of the third edition of *How to Teach Your Baby to Read* that reads "*This book is respectfully dedicated to my wife, Hazel Doman, who through their mothers, has taught hundreds of one- two- and three-year-old brain-injured children to enjoy reading*"[328] seems to carry certain thinly veiled claims.

In the light of what we have learned from years of research in developmental psychology, as well as from everyday observations, by age one, the child is able to say one or two words, responds to its name, imitates familiar sounds, and can follow simple instructions. Between the age of one and two, the vocabulary size reaches 5-20 words, the child is able to say two-word sentences and to express their wishes by saying words like "more" or "up." They understand the word "no." In its second and third years, the child is able to refer to itself as "me," put nouns and verbs together to utter short (logical/coherent) phrases, and their vocabulary increases to approximately 450 words. The child can form some simple plurals and is able to answer "where" questions.

There are at least three revolutionary claims concealed in this highly appreciative dedication. They are all set in the context of child development. The first revolutionary presupposition is that children can actually read before they start to talk. Such a claim, it goes without saying, pulverizes both research results and everyday observations to the point where they are simply worthless. The second subversive claim one may identify is that a brain-injured child could develop faster than typical non-injured children. And last but not least, a third inference can be made that the brain-injured children's mental processes are so advanced that not only can they read, but they also "enjoy" reading! In all fairness, there is a really exciting methodological brain teaser embedded in these claims. How can we measure the ability to read and the degree of enjoyment among children who cannot speak?

It is no surprise, then, that the scientific profession revealed an intense interest in the examination of such a promising concept soon after its rumored sensational effects had first been reported. Having conducted a thorough analysis of the evidence, in 1968 the American Academy of Pediatrics and the American Academy of Neurology issued a joint cautionary statement, in which they advised physicians who make decisions and formulate recommendations with regard to treatment and therapy for neurologically impaired children that they should be aware no hard evidence had been collected to date for the efficacy of either the Doman-Delacato method or the related treatment programs. They also emphasized the need for research conducted by acknowledged experts based on well-constructed control groups.[329]

As might have been expected, the first research studies were published soon after and they intensified the skeptical approach demonstrated by scientific institutions. Distinct findings were obtained by Sarah Sparrow and Edward Zigler in 1978. Forty-five children with severe mental impairments participated in their study. The researchers put them into three groups of fifteen participants each. Children from the first group received treatment designed in compliance with the Doman-Delacato method for two hours a day, five days a week, over the course of one year. Within the same length of time, children from the control group underwent equally intense activities with their grandparents. These were motivational sessions aimed at creating positive, success-oriented interactions to raise self-esteem and bring about a sense of self-efficacy. Participants in the third group received standard institutionalized care. To be able to determine how effective each of the therapeutic procedures was, before the commencement of the project the investigators had employed a number of evaluative measures, such as the IAHP Developmental Profile, IQ tests, motor and language development scales, as well as measurements of emotional, social, and maladaptive behaviors. After a year, the children were re-evaluated. Unfortunately, in the majority of the measures no post-test differences among the three groups were found. Yet – what is important! – all three groups showed some performance improvement. The researchers concluded that there was no empirical evidence for the analyzed method to be recommended as an efficacious rehabilitation therapy for children with severe mental impairment.[330]

Similar poor results were obtained when examining reading progress in children who participated in development programs offered by the Institutes.[331] Some of those reported by their therapists to be "miraculously cured" proved to be children with an earlier misdiagnosis or an overly pessimistic prognosis. Other (very well designed) research projects cofounded by state-owned and private institutions were abandoned after the Institutes suddenly withdrew their previously confirmed participation.[332] It additionally turned out that the diagnostic tool used in the Institutes – the Developmental Profile – had never been subjected to standardization.

The studies referred to above constitute just a small fraction of the entire research effort that continued to bring conclusive findings showing the method's lack of efficacy. They all led to a subsequent statement by the American Academy of Pediatrics in 1982 that concluded with the following: "Based on past and current analyses, studies, and reports, we must conclude that patterning treatment offers no special merit, that the claims of its advocates are unproven, and that the demands on families are so great that in some cases there may be harm in its use."[333]

To a person with a rational mindset, the above should have been a sufficient indication as to the reliability of the method – but apparently not to the Institutes. They continued to grow, while their heralds of good hope set parents to backbreaking work, all the while draining money from their pockets. Around that time, scientists persisted in publishing critical papers that left no shadow of doubt as to the method's lack of efficacy. Concurrent IAHP publications, available outside the official scientific press, reported new cases of "miraculous healings." This situation elicited another firm statement from the American Academy of Pediatrics in 1999. Their conclusions were identical to the ones communicated in 1982; however, this time they were expressed far more forcefully. The Academy stated, among other things, that:

> Treatment programs that offer patterning remain unfounded; i.e., they are based on oversimplified theories, are claimed to be effective for a variety of unrelated conditions, and are supported by case reports or anecdotal data and not by carefully designed research studies. In most cases, improvement observed in patients undergoing this method of treatment can be accounted for based on growth and development, the intensive practice of certain isolated skills, or the nonspecific effects of intensive stimulation. ... The demands and expectations placed on families are so great that in some cases their financial resources may be depleted substantially and parental and sibling relationships could be stressed.[334]

The same statement was reiterated by the Academy in 2002, and once again in 2005.

The United Sates was not the only country where scientists protested against engaging parents and children in tortuous and worthless therapy. A serious response to the staggering success of the Doman-Delacato method came with a statement issued by a team of French investigators who explicitly defined the therapy as a moral deceit. This standpoint was formulated indirectly as a result of strong pressure exerted by parents who had used the method and demanded reimbursement for all treatment costs they had incurred. The French Ministry of Social Affairs and Employment (*Ministre des Affaires sociales et de l'Emploi*) entrusted a team of experts with the task of conducting a detailed analysis of the Doman-Delacato theory. The team, led by Stanislaw Tomkiewicz, a child neurologist from Paris, investigated the therapy for eighteen months. Their findings were officially published in 1987 in the report *Méthode Doman Evaluation*.[335]

The document includes interviews, accounts of observations of children treated with the Doman-Delacato method, reports, surveys, and results of statistical analyses. The research involved 249 children from a number of countries. Once again, despite methodological problems combined with the emotionally-charged atmosphere surrounding the method itself, the published results showed in a clear-cut manner that there was no scientific basis to deem the method efficacious in the treatment of developmental disorders. No case of a child healed through the therapy was found. The authors of the report's significant conclusion was that the method was undoubtedly too expensive when compared to the effects that should possibly be expected of it. Therefore, it was further stipulated that under no circumstances should the therapy be funded by the state social security system. Due to its deceitful character that aroused false hopes, it was also referred to as a moral deception. Notwithstanding all these concerns, the report did advise against prohibiting the use of the method, out of fear that it might cause an effect opposite to the one desired.[336]

The Doman-Delacato concept makes use of assumptions that contradict the knowledge we have today on the growth of the brain and the nervous system. At present, even the most adamant skeptics do not deny the role genetic factors play in the development of intelligence. It is widely accepted that genetic-based differences within the population account for about 50% of variability in our intelligence.[337] Claims about children's unlimited developmental potential are nothing more than a simple fraud. The authors and advocates of the method in question additionally ignore an obvious fact, i.e. that various forms of mental impairment have diverse causes and origins. While, for instance, Down syndrome is a disorder of genetic origin, the Institutes proposes treating it with the same therapy as cerebral palsy. By the same token, if the therapy is based on the Doman-Delacato method, the child is forced into a regimen that has nothing in common with the natural rhythm of existence. For most of the time, the baby patient is exposed to an array of stimulation forms, and, as it stays completely passive throughout, is additionally prevented from gaining a sense of independence and empowerment. An individually tailored teaching program is repeated from eight to twenty times a day, for two or three minutes. A diet low in sugar and water is kept and, consequently, salt and sweets are eliminated. Every day the child is fed with unsweetened fruit juices and is given a lot of vegetables and lean meat. The fixed therapeutic program is rerun from several to a dozen times a day. Sometimes the child must go through a meticulously planned twelve-hour regimen. Exercises are carried out seven days a week. No orthopedic aids, surgery, or pharmacological treatment are recommended in this method. Instead, the strong and ongoing commitment of parents and volunteers is required, not to mention a sufficiently long period of time for them to carry out the rehabilitation program.[338]

The proposition to organize exercises so that they are performed every few minutes and for twelve to fourteen hours a day does not allow for the fact that the mobility rate in children with brain damage may be up to twenty times lower than in non-impaired ones. This limitation excludes efficient movement control. The Doman learning technique that draws upon, among other theories, classical conditioning, passes over fundamental laws of the child's psychological development and is in fact closer to animal training. Children who are trained with the Doman approach demonstrate complete helplessness in a broad range of simple life circumstances, caused by the lack of activity and limited opportunity for spontaneous development of skills. Besides, even apart from the question of considerable costs of the "treatment program," it is exceptionally unethical that the entire burden of responsibility for effects falls onto parents, who would frequently develop a feeling of guilt if confronted with therapeutic failures. It is they who are blamed for the lack of progress, despite the fact that the development of children is determined, to a large extent, by the nature of the brain damage. On the other hand, if the child shows improvement, it is hailed as a success of the therapy and inevitably the method itself. Hence, there is only either a lack of improvement and depression along with parental guilt, or positive effects and an unfounded belief in the effectiveness of the treatment. The burden that parents must bear is well illustrated by the following passage from a book by Charles Hart:

> In 1981, Dr. Edward Zigler, professor of psychology at Yale, complained that Doman had raised unrealistic hopes in many parents, but added, "I know of no accounts relating the experiences of families sometime after they have assayed or completed the treatment."
> Eight years later, Zigler had his chance. Berneen Bratt invited him to write the introduction to her book, *No Time for Jello*, in which she describes years of stress and fatigue, following the program she thought would help a son with cerebral palsy. Her experience resembles that of other parents, including many who have a child with autism. None of these families can claim a cure, in spite of incredible effort, commitment, and inconvenience.
> "This is a cautionary tale showing how and why a middle-class educated family could fall victim to a therapy that doesn't work," Zigler writes in the introduc-

tion. "The book powerfully evokes the emotional ups and downs associated with any therapy."

In *Mixed Blessings*, Barbara and William Christopher share their experience with the Doman program. They explain that the staff told them their seven-year-old son should be allowed to walk only one hour a day. The rest of the time he had to crawl and follow an exhausting schedule that included breathing exercises, patterning, and studying flash cards prepared by his mother.

The Christophers invested tremendous amounts of time, energy, and money in this program, but after three years, they lost hope and began looking for a residential program for their son. The parents finally burned out, and their son's behavior indicates he may have felt that way too.[339]

The method and its supporters have fallen into a situation that is well conveyed by the notion of the "besieged fortress syndrome," which is, as a matter of fact, quite commonly seen in the pseudosciences. The IAHP does not admit those who have a critical opinion of the propagated method. Their founder himself, bandying catchy phrases about "novel ideas" and progress, attempted to discredit representatives of science by dubbing them as conservatives and implying they were fearful of innovations. In his statements he used eristic tricks to disparage scientists and promoted his own line of reasoning as the only alternative to follow: "We discovered that it mattered very little (except from a research point of view) whether a child had incurred his injury prenatally, at the instant of birth, or postnatally. This was rather like being concerned about whether a child had been hit by an automobile before noon, at noon or after noon. What really mattered was which part of his brain had been hurt, how much it had been hurt, and what might be done about it.[340]

If such a message reaches parents who do not have much in common with science, it is indeed sufficiently persuasive to shape an image of scientists as individuals wasting their time on useless matters. It is simultaneously implied that throughout this time, Glenn Doman himself, with the help of his assistants, made "a fundamental discovery" and proved that for the sake of treatment it is essential to identify which part of the brain has been damaged!

Truly, one must really assume the recipients to be totally ignorant if they are expected to believe such nonsense. And nonsense is just the thing with which Doman lavished upon those who wanted to listen to him.

Experts who managed to sneak into the Institutes confirm that Doman was considered a true guru. One of his most frequently communicated messages entailed extreme criticism of physicians, whom he accused of having a supposedly negative approach towards children with disabilities. This well thought-out propaganda aims to form a negative mindset towards mainstream medicine among parents, and, at the same time, encourages them to use the support provided by the Institutes. Also in this respect Doman did not hold back from spouting simplifications. If truth be told, he was not too original in doing so. One of the eristic tactics that he continued to use is a frequent tool in debates in which one intends to discredit a given treatment approach. It involves convincing the other party that the questioned method is focused solely on treating symptoms: "In those early days, the world that dealt with brain-injured children held the view that the problems of these children might be solved by treating the symptoms which existed in the ears, eyes, nose, mouth, chest, shoulders, elbows, wrist, fingers, hips, knees, ankles and toes. A large portion of the world still believes this today."[341]

Obviously, we will not learn from the author of this statement which "groups" specifically he had in mind. Indeed, throughout his whole book Doman cites fewer sources than I do in this one section alone. If, however, his readers and audiences do not bother to question the sources and true availability of a particular treatment, they may find his argumentation credible.

In the Institutes parents receive proper training and more. They can also buy complementary rehabilitation accessories, vitamin supplements, nutritional preparations and other things that are indicated as necessary during the therapy, all naturally sold at appropriately high prices. Some cases of "miraculous healing" are easily understood, as Glenn Doman himself suggests how they should be explained:

> At The Institutes we see more than a thousand brain-injured children each year. There is hardly anything a mother fears more than having a brain-injured child. And if she suspects it, she wants to find out at the first possible moment so she can start doing immediately whatever has to be done.
> In over nine hundred out of a thousand cases seen at The Institutes, it was Mother who first decided that *something* was wrong with her baby. In most cases Mother had a very difficult time convincing anyone--including the family doctor and other professional people—that something was wrong and that something should be done about it at that instant.
> No matter how hard or how long everyone tries to talk to her out of it she persists until the situation is recognized. Sometimes it takes her years. The more she loves her baby, the more detached she makes herself in evaluating its condition. If the child has a problem, she will not rest until it is solved.
> At The Institutes we have learned to listen to mothers.[342]

There is no better grounds to get a psycho-business going. Here is a desperate mother with an imaginary conviction about her daughter or son's disability who has been seeking understanding from professionals. She has repeatedly been turned down by physicians and told that there is nothing wrong with her child. It is only the IAHP therapists who, having learned to "listen to mothers," confirm her "diagnosis" and suggest starting the therapy. In this way the woman finds both comfort and confirmation of her own worst suspicions, as well as support, i.e. reassurance that every disability can be cured. This is a fear-and-relief manipulation technique in its classic form! Any therapy that is provided to a child diagnosed in this manner must end up in success. Children's development is a natural process and, unsurprisingly, it is not easy to disrupt it. Intensive and systematic care, as long as it does not interfere with the process, must contribute to improvement.

In Western Europe, the Doman-Delacato approach made its debut in the latter part of the 1970s. One of the first centers to promote the method was an institute in Bridgewater near Bristol, England. Having come into conflict with Doman, it split and started to operate on an independent basis. In Europe, Glenn Doman's Institutes experienced their full bloom in the 1980s. Due to a lack of expected effects, the initial enthusiasm of parents would fade away after several years of having gone through grueling exercise regimens. In Western European countries, it was also the unequivocally critical attitude of scientists that contributed to an evident slump in the popularity of the Doman-Delacato method, followed by closures of some centers due to a failure to attract new clients. The method reached Eastern Europe on the rising tide of political transformations in the early 1990s, and was promoted through the agency of the Bridgewater Institute branches.[343] Costs of training of the Institute's licensed therapists were partially covered from public funds. At present, in Poland the Doman-Delacato method is quite commonly used and recognized as a conventional treatment method. I have not managed to find either a single critical analysis of the approach in Polish or a statement issued by a board of Polish scientists, physicians, or therapists. Can it be that criticism of the method is unable to keep pace with its apologists?

Without straining myself too much though, I have succeeded in identifying a whole host of institutions and individuals who put the method into practice, along with numerous cases of providing support and funding for the method by public bodies. Shamefully, Polish scientists also write about the method without a hint of criticism. Such publications are authored by, for instance, academic workers of the Institute of Psychology at the University of Wrocław[344] or the University of Physical Education in Wrocław.[345] The method has also become a part of the official academic curricula, e.g. at the Medical University of Silesia (at the physical therapy faculty), the University of Gdańsk (at the faculty of education), and the Maria Curie Skłodowska University in Lublin (the faculty of clinical speech therapy). It is additionally available in child care centers, psychological and educational counselling centers, and countless other institutions that provide rehabilitation and therapy-related services.

I think the time has come to look at the gray-haired, noble figure of Glenn Doman and raise a few additional questions about his true motives. Did he adopt those ill-founded assumptions of his theory driven by a concern for the welfare of children? Was it the same concern that guided him when he continued to ignore statements issued by American pediatricians? Did physicians who issued successive statements questioning the basis and practices of his method for nearly 40 years do this out of envy or because they were blinded by other reprehensible purposes? Is it for the welfare of children that the

IAHP withdrew from a research program whose aim was to conduct a reliable verification of their methods? Is Glenn Doman's much-acclaimed *gentle revolution* not, by any chance, about entangling parents in a long-standing, torturous effort that brings no results, but is fueled by the growing feeling of guilt and remorse?

Glenn Doman dispels all such doubts with one simple piece of advice: "If, after all this time and this wealth of experience, a parent should ask for some important advice in a very short sentence, that advice would be: *Go joyously, go like a wind and don't test.*"[346]

---

[320] J. Traub, "Goodbye, Dr. Spock: Vignettes from the Brave new World of the Better Baby," *Harpers Magazine*, (March, 1986): 57-64, 58.

[321] G. Doman, *How to Teach your Baby to Read. The Gentle Revolution* (Pennsylvania: The Better Baby Press, 1990), xv.

[322] R. J. Doman, E. B. Spitz, E. Zucman, C. H. Delacato, and G. Doman, "Children with Severe Brain Injuries. Neurological Organization in Terms of Mobility," *Journal of the American Medical Association 174*, (1960): 257–262.

[323] G. Doman, *What to do About your Brain-Injured Child* (New York: Doubleday & Co, 1974).

[324] T. Cadden, "Award to Glenn Doman, Founder of The Institutes for the Achievement of Human Potential, Author, Humanitarian," *Market Wired*, (November 2, 2007): http://www.marketwired.com/press-release/award-glenn-doman-founder-the-institutes-achievement-human-potential-author-humanitarian-788060.htm

[325] G. Doman, and R. Peligra, "Ictogenesis: the Origin of Seizures in Humans. A new Look at an old Theory," *Medical Hypotheses 10*, (2003): 129-132.

[326] J. Dziczkowski, and E. Zieliński, *Metody Rehabilitacji – Metoda Glenna Domana*. Specjalny Ośrodek Szkolno-Wychowawczy w Szklarach Górnych, 2010, http://www.sosw-szklarygorne.com/index2.html.

[327] Doman, *How to Teach*, vii.

[328] Ibid., v.

[329] Committee on Children with Disabilities, American Academy of Pediatrics; P. R. Ziring, D. Brazdziunas, W. C. Cooley, et al. "The Treatment of Neurologically Impaired Children Using Patterning," *Pediatrics 104*, (1999): 1149–51; American Academy of Pediatrics. "Doman-Delacato Treatment of Neurologically Handicapped Children," *AAP Newsletter*, (1 June, 1968): http://pediatrics.aappublications.org/content/104/5/1149.full

[330] S. Sparrow, and E. Zigler, "Evaluation of a Patterning Treatment for Retarded Children," *Pediatrics 62*, (1978): 137-150.

[331] M. P. Robbins, "Test of the Doman-Delacato Rationale with Retarded Readers," *Journal of the American Medical Association 202*, (1967): 389-393.

[332] B. S. Rosner, *Final Report on Planning Grant: Treatment of Brain-Injured Children*. Presentation for Vocational Rehabilitation Administration, National Association for Retarded Children, Given Foundation, 1967. As cited in: American Academy of Pediatrics, "The Doman-Delacato Treatment of Neurologically Handicapped Children," *Pediatrics 70*, (1982) 810-812.

[333] American Academy of Pediatrics, "The Doman-Delacato," 811.

[334] American Academy of Pediatrics, "The Treatment of Neurologically Impaired Children Using Patterning," *Pediatrics 104*, (1999): 1149–1151.
[335] S. Tomkiewicz, D. Annequin, and I. Kemlin, *Méthode Doman Evaluation* (Paris: CTNERHI, 1987).
[336] Ibid.
[337] This is a very rough and "safe" figure. Radical adherents of genetically determined differences in intelligence estimate it even at over 80%. For a review see: R. Plomin, J. C. DeFries, G. E. McClearn, et al., ed., *Behavioral Genetics (4th ed.)* (New York: Worth Publishers, 2001).
[338] Tomkiewicz et al., *Méthode Doman Evaluation*.
[339] C. Hart, *A Parent's Guide To Autism: Answers to the Most Common Questions* (New York: Simon and Shuster, 1993), 133-134.
[340] Doman, *How to Teach*, xiv.
[341] Ibid., *xv*.
[342] Ibid., 86-87.
[343] E.g. http://fundacja.dajszanse.free.ngo.pl/index.php
[344] D. Łupicka-Szczęśnik, „Metoda Domana-Delacato jako Metoda Psychokorekcyjna i Psychostymulacyjna," In *Wspomaganie Rozwoju - Psychostymulacja, Psychokorekcja*, ed. B. Kaja (Bydgoszcz: WSP, 1997), 231-238.
[345] T. Bober, and K. Kobel-Buys, ed. "Mózgowe Porażenie Dziecięce. Z Doświadczeń 3-Letniego Programu Rehabilitacyjnego," *Studia i Monografie 81*, (Wrocław: Wydawnictwo AWF, 2006).
[346] Doman, *How to Teach*, ix.

CHAPTER 12: GO TO SCHOOL, GET A TASTE OF PSEUDOSCIENCE: EDUCATIONAL KINESIOLOGY

The story of another famous therapy inventor begins like a moving Hollywood feature. An American boy named Paul suffers from dyslexia. Fed up with struggling at school and being the butt of classroom jokes, Paul takes matters into his own hands. He begins experimenting with drills of his own invention. Step by step, he develops a program that helps him master and, eventually, overcome dyslexia. The boy's name is Paul Dennison and his eponymous method is also known as educational kinesiology or *Brain Gym®*. And this is just the beginning of this edifying story.

As we saw in the previous chapter, the Doman-Delacato method expanded into a smoothly running psycho-business. Its success, however, pales in comparison to a veritable empire built on educational kinesiology. While the proliferation of the Doman-Delacato method capitalized on the assistance of public institutions, its creators and promoters even receiving national honors, educational kinesiology has achieved a feat that seems to have eluded Doman-Delacato: securing direct backing from official government-funded programs. Not content with its original area of child rehabilitation, educational kinesiology succeeded at making inroads into the world of early education. Its remarkable resilience allowed educational kinesiology to become one of those pseudoscience disciplines that thrive and even flourish despite popular criticism. Indeed, according to data quoted by Brain Gym® International, the method is present in eighty-seven countries worldwide and its tenets have been translated into forty different languages.[347]

Although Brain Gym® is used in learning environments every day around the world, we do not have comprehensive statistical data showing its reach. "According to a training schedule published on the Official Brain Gym® Web site, there were 337 different trainings scheduled between February and December of 2006. Of those trainings, 211 were scheduled in the United States, with the remaining 126 trainings scheduled in Canada, Australia, the United Kingdom, France, Germany, Switzerland, Belgium, Japan, Indonesia, and Singapore."[348]

The Brain Gym® International website provides an overview of their current, quite impressive, offering. In 2015, there were 136 training courses scheduled in the United States, twenty in Canada, fifteen in Spain, seven in China, six in Hungary, five in Austria, four in Australia, three in the UK, and one in Slovenia. Together this made 207 training courses organized by Brain Gym® International alone.[349]

Ben Goldacre, an established columnist at *The Guardian* and a well-informed critic of the method said: "I've seen the 12,000 Google hits for Brain Gym® on UK government web pages."[350] He concluded that "Brain Gym® is an incredibly popular technique, in at least hundreds of British state schools, promoted all over government websites, and with a scientific explanatory framework that is barkingly out to lunch."[351]

Little data is available about the popularity of the concept in the USA, but a random sample of accounts suggests it exists on a similar level to that in the UK. Joan Spalding, a Brain Gym® popularizer, stated on her website: "As part of the Supplemental Educational Services Program for the state of Colorado, I have developed an after school program called SOAR (Students Organizing and Reaching Goals). This program incorporates the Multiple Intelligences theory and Brain Gym® exercises in an experiential language arts and mathematics enrichment program."[352]

In Poland, educational kinesiology has established a power base mainly in pre-school and early school education and in psychological and pedagogical therapy. The method has been given full blessing by the Ministry of Education. In 2005, an educational kinesiology-based curriculum for six year olds[353] was certified for use in pre-school education.[354] This is sufficient for thousands of children in my country to start their day with basic exercises recommended by educational kinesiology. Typically, this involves making the so-called lazy eights, or tracing the shape of an eight turned on its side (the sign of infinity) using just a hand or an aid. This exercise is purported to improve visual coordination and peripheral vision, "activate the brain to cross the visual center line, integrate the brain hemispheres, improve the mechanism responsible for reading and character recognition in writing and support reading comprehension."

The Polish Ministry of Education is not the only body with a substantial role in spreading kinesiology. A list of organizations promoting and financing educational kinesiology in Poland is very long indeed. In 2006, a large-scale learning campaign was launched under the patronage of the Polish National Commission for UNESCO. One of its objectives was "to promote accelerated learning," which the noble institution's official online magazine defines in a way that contains some rather surprising elements:

> "Accelerated learning" is an umbrella term which combines scientific data coming from four main areas of learning about the functioning of the brain and learning processes, including:
> - Neurological study of the brain;
> - Theory of multiple intelligences;
> - Neurolinguistic Programming (NLP);
> - Educational kinesiology.[355]

There you go.

However, educational kinesiology would not have qualified as proper pseudoscience without having penetrated into the academic world as well. The previously mentioned Joan Spalding has this to say about herself: "I have recently completed a Ph.D. doctoral program in Educational Leadership at

Colorado State University. My recent doctoral dissertation focused on the use of the Brain Gym® exercises in the classroom as it relates to learning and achievement." And she adds "As an educational consultant, for the past 19 years, I have taught classes for Colorado State University in Educational Kinesiology." Her achievements in the area of popularization of educational kinesiology among academic teachers stretch further, as she adds: "I have been a member of the Rotary International organization since 1990. In 1991, I volunteered for a one year program to teach Brain Gym® at universities for teachers and psychologists in Russia, Latvia and Tajikistan."[356]

In Poland educational kinesiology has been taught at several public universities. Alongside the obvious choices of psychology and pedagogy, the concept has established itself in rehabilitation studies and is also present in medical university curricula.

Paul Dennison developed his method in the late 1970s and conducted his first educational kinesiology workshop in 1981. For a surprisingly long time, his operation enjoyed a criticism-free ride save for some of the doubts raised over the Doman-Delacato method, which Dennison drew on for some of his theoretical foundations. It was not until 1991 that the first questions were asked about educational kinesiology, and these did not even come from scientists. The German consumer organization *Stiftung Warentest*, a provider of test ratings of goods and services, took aim at educational kinesiology. A study commissioned by the organization yielded a rather straightforward result: the muscle test, an inherent component of the method and the basis for initial diagnoses, was found to be worthless. From that time onward, *Stiftung Warentest* has recommended that consumers avoid services involving the muscle test.[357]

Among the first within the scientific community worldwide to speak out about the usefulness of educational kinesiology were allergologists. The European Academy of Allergy and Clinical Immunology published its official position on the matter, concluding that there was no documented rational evidence for the diagnostic efficacy of applied kinesiology. The method was not recommended for clinical practice, as it failed to show any difference from a placebo in studies and its adverse effects could not be ruled out.[358]

In 1996, the German Land of Schleswig-Holstein officially prohibited the muscle test and energizing exercises offered by educational kinesiology from being integrated in the Land's school curricula. The Land's Ministerium für Bildung Wissenschaft, Forschung und Kultur based this decision, which also included a negative opinion about neurolinguistic programming, on research published in the same year.[359]

More criticism emanated from German academia in the following year. Professor of Special Pedagogy Karl-Ernst Ackermann wrote, in what was an official position of his FernUniversitaet, Hagen, that rather than therapy, educational kinesiology was a commercial product and that no available scientific research could be found to plausibly identify it as a therapy.[360] In the same year, another researcher, Christoph Kant, picked out errors in the theoretical foundations cited by Dennison.[361] A year later, Professor of Psychology Erwin Breitenbach confirmed that the muscle test, the fulcrum of Dennison's therapy, was useless from a scientific perspective and recommended against employing educational kinesiology techniques in schools and pre-schools.[362]

In 2001, a German federal cross-ministerial committee for combating cults, Ständiger Interministerieller Ausschuss Zur Bekämpfung Von Sekten, probed the educational kinesiology movement. It published a report concluding that educational kinesiology's modus operandi was similar to that of a sect. The committee also stated that after completing paid courses individuals received support to expand their groups of customers, which was ostensibly intended to generate material gains through exploitation of desperate people who were seeking help.[363]

The year 2001 also saw the first critical study on educational kinesiology published in English. In her doctoral thesis devoted to one of Dennison's techniques known as Brain Gym®, the author conducted her own study that demonstrated no significant differences between results of students using the Brain Gym® technique and the control group.[364]

Subsequent years saw a barrage of criticism leveled at educational kinesiology. The previously quoted Ben Goldacre, a medical doctor and well-known science journalist, launched his campaign in 2003. In his kick-off article on Dennison's method Goldacre says "On the off chance that it might not be rubbish I looked it up on the main public research databases. Nothing supported their assertions."[365]

Goldacre mentions kinesiology a number of times, including:

> I've accidentally stumbled upon a vast empire of pseudoscience being peddled in hundreds of state schools up and down the country.[366]

> The science they use to justify this so often seems to be bogus, empty PR, that promotes basic scientific misunderstandings, and most of all is completely superfluous in every sense except the commercial: because the ropey promotional "science" is the cornerstone of their commercial operation, they need it to promote themselves as experts selling a product that is unique and distinct from the obvious, sensible diet and exercise advice that you can't copyright.[367]

While Goldacre's media campaign continued, extensive scientific research evaluating the scientific plausibility of educational kinesiology was slower to come, but when it started, it did so simultaneously in Poland and the UK. In May 2006, Professor Usha Goswami, the head of the Centre for Neuroscience in Education at the University of Cambridge, published a paper in *Nature* where she concluded that Brain Gym® and similar programs were based on myths, not facts, and should be eliminated immediately.[368] At nearly the same time, i.e. in October 2006, the Committee of Neurobiology of the Polish Academy of Sciences published a study on educational kinesiology concluding that:

1. The tenets of Dennison's method are incompatible with modern knowledge on brain functioning;
2. Most of the claims about the results of purported scientific research on which the method is based are false;
3. Descriptions of the processes and principles of brain functioning published in educational kinesiology studies make no sense from a scientific perspective;
4. Dr Paul E. Dennison probably never carried out scientific research about the influence of exercises he proposed on brain processes and on the results of learning.[369]

2007 saw even more critical studies, when data from neurosciences were analyzed to conclude that Brain Gym® was purely a non-scientific procedure.[370] Keith Hyatt published a paper evaluating the theoretical foundations and a review of available studies on the effectiveness of the brain gym method. He summed up the section on the theoretical foundations as follows:

> Brain Gym® materials and writings have consistently promoted the notion that the exercises "activate the brain" and facilitate "whole brain learning." However, Brain Gym® materials provide no scientific support that the brain needs to be "turned on," nor do the materials provide research support regarding whole brain learning or even a definition of exactly what this term means. These terms appear to be phrases that capture the imagination and lead the uninformed reader to believe in something for which there is no theoretical support. In fact, research findings have strongly refuted the theoretical foundations on which Brain Gym® was developed. Neurological patterning has been described as fraudulent, cerebral dominance has not been linked to learning, and perceptual-motor training has not withstood rigorous scientific investigation.[371]

Sadly, an equally negative verdict was produced by a review of available research:

> Taken together, these studies clearly failed to support claims that Brain Gym® movements were effective interventions for academic learning. They were overcome by methodological difficulties, and two studies failed to address academic learning at all. ... the majority of the articles listed were not reviewed, because quality research should be published in peer-reviewed journals available through academic libraries rather than sold by the organization promoting the treatment or publishing the journals, as was the case with Brain Gym®.[372]

In autumn of the same year, a conference was held in Warsaw under a telling title: *Educational Kinesiology. Science, Pseudoscience or Manipulation?* It was attended by sociologists, psychologists, special pedagogy experts, neurophysiologists, and pediatricians, as well as experienced therapists, rehabilitation practitioners, and surgeons. The published volume of conference proceeds[373] is unparalleled even when compared to literature available in the English language, and for this reason some of the conclusions contained therein seem worthy of publishing to provide a complete view of the educational kinesiology landscape. Let us first hear from a neurophysiologist. Professor of neurophysiology Anna Grabowska of the Nencki Institute of Experimental Biology thus reports her evaluation: "All the papers that I have read, and there were many, including some sent to me directly by the International Institute of Neurokinesiology, contain propositions which are false from the perspective of contemporary science and use unacceptable terminology of a kind that reveals that the authors are utter dilettantes when it comes to brain functioning. It also seems to me that the citing of such genuine authorities in the area as A. Damasio, E. Goldberg, R. Sperry and J. Piaget in order to add credence to educational kinesiology is a clear case of misuse."[374]

Maria Borkowska, a pediatrician and expert on rehabilitation, concluded her intervention by saying: "This short review of educational kinesiology fundamentals and their comparison with current medical knowledge places the former in a position of a method lacking any sound theoretical basis in the areas of neurophysiology and neuropsychology. It represents a quasimedical and quasipsychological approach."[375]

One of Poland's greatest authorities on autism, Professor Ewa Pisula from the Department of Rehabilitation Psychology at the Faculty of Psychology of Warsaw University, said: "In summary, we must conclude that an application of educational kinesiology to autistic children has been found to be fundamentally lacking. Modern psychology and neurobiology offer no arguments that would support the claims made by the creators of this method. ... It is worth pointing out that much more can be done for a child suffering from autism than massaging its 'thinking points' or training it to trace 'lazy eights.'"[376]

Finally, Neuropediatrician Zofia Kułakowska summed it up nicely:

Reservations about Dennison's educational kinesiology:
scientific: missing neuropsychological diagnosis;
- dubious interpretation of reflexes;
- mistaken reports on the function of disconnected brain hemispheres;
- missing adaptation to child's age in proposed therapy;
- references to Eastern philosophies – fragmentary and unexplained;
- mental manipulation of parents and educators.

social:
- high prices;
- misleading educators.[377]

This is just a small sample of a range of similar conclusions found in this volume, which is generally in line with studies published in German and English. From a scientific perspective, however, this should be sufficient for readers to make up their minds about the merits of educational kinesiology. Perhaps, however, they may also choose to take a look at the workings of Denninson's business.

The reader will have noticed the ® sign cropping up throughout this chapter. It is not here by accident, or to test the reader's observation skills. This symbol means that the designated brand has been officially registered in a commercial register and is protected. The symbol also conveys a warning: nobody is allowed to use this brand – unless a certain fee is paid. Now you might want to think about these 87 countries and the dozens or perhaps hundreds of organizations in each of them using this protected method. Each one generates a little stream of cash that trickles down to meet with others, and eventually they join together in one river that runs into Brain Gym® International. If you expected that Paul Dennison could be found in this *non-profit* organization established in 1987 you were mistaken. Indeed, Brain Gym® International is owned by another *non-profit* organization, The Educational Kinesiology Foundation of Ventura, California. Only there will we find the boy who had a reading problem and was ridiculed by his peers, but who finally took matters in his own hands...

Brain Gym® International is not the only source of funds channeled into the Educational Kinesiology Foundation. More cash flows from Edu-Kinesthetics Inc., a management company responsible for the rights to everything involved in educational kinesiology, from books, to posters, to music, to exercise aids, to all kinds of gadgets that can be sold as a complement to the therapy. Unfortunately, I lack the kind of investigative journalist talent that would be required to properly estimate the revenues of Denninson's foundation. I shall leave this task to the imagination of the reader and to the more inquisitive truth-seekers out there.

Now, what is it exactly that Denninson chose to protect when he registered his brand? Contrary to what might be expected, it is not his rubbish invention, but the ancient recommendations of Ayurveda masters. While Dennison invariably claims that everything he proposes is based on neurophysiology,[378] many exercises included in educational kinesiologists appear to be identical to recommendations of traditional Hindu medicine! Some of these are mudras (hand gestures) intended to bring energy and harmonize the workings of the mind, while others are supposed to have an effect on meridians. Naturally, none of these remedies has been empirically tested for its efficacy, and certainly not by neurophysiologists, as claimed by Denninson. Even more interestingly, the inventor of educational kinesiology keeps the origins of his "innovative" brain gym method shrouded in secrecy. While some references to these origins could be found in the original editions of his books, they gradually disappeared from subsequent editions.[379] It would be difficult to tell whether this approach was dictated by accusations of plagiarism, to make sure the method appeared "unique," or by something else altogether. It is a shame that representatives of UNESCO, Rotary International, and other organizations financing the promotion of educational kinesiology fail to realize that what they are really promoting are methods derived directly from religious practices of the Far East and based on a religious perception of reality.

Critical minds among us can find some solace from a continually expanding body of publications about educational kinesiology released by both critics of the method and its promoters. They provide ample material to study the operation of meridians, mudras, the flow of cosmic energy, the use of the muscle test to strengthen communication with others, and other similar techniques. Should you, the reader, have found educational kinesiology fascinating and regard my account as incomplete or unreliable I would strongly advise you turn to these sources for a better understanding of the matter.

---

[347] http://www.braingym.org/about

[348] K. J. Hyatt, "Brain Gym® Building Stronger Brains or Wishful Thinking?" *Remedial and Special Education*, 28, (2007): 117–124.

[349] http://www.braingym.org/schedule

[350] B. Goldacre, "Brain Gym® Exercises do Pupils no Favours," *The Guardian*, (18 March, 2006): http://www.guardian.co.uk/life/badscience/story/0,,1733683,00.html

[351] B. Goldacre, "Exercise the Brain Without this Transparent Nonsense," *The Guardian*, (25 March, 2006):
http://www.guardian.co.uk/life/badscience/story/0,,1739365,00.html.

[352] http://www.mindsync-us.com/about.html

[353] A. Boniecka, A. Kozyra, M. Wypchło, *Mój Kuferek. Program Wychowania i Kształcenia dla Sześciolatków* (Warszawa: JUKA-91, 2005).

[354] Certificate DKOS-5002-11/05.

[355] *Internetowy Magazyn TRENDY*, 1, (2006): 11,
http://bc.codn.edu.pl/Content/36/2006_01_trendy.pdf

[356] http://www.mindsync-us.com/about.html

[357] K. Rychetsky, „Kinezjologia Edukacyjna w Europejskiej Perspektywie," in *Kinezjologia Edukacyjna: Nauka, Pseudonauka czy Manipulacja?* ed. K. Korab (Warszawa: Instytut Badań Edukacyjnych, 2008), 133-136.

[358] *Alergia Astma Immunologia*, 3, (1994): 152.

[359] Ministerium für Bildung Wissenschaft, Forschung und Kultur des Landes Schleswig-Holstein, *Einsatz von Methoden der Kinesiologie und des Neurolinguistischen Programmierens (NLP) in Schulen und Lehrerforbildung*, 1996, http://www.schulrecht-sh.de/texte/k/kinesiologie.htm

[360] K. E. Ackermann, *Stellungen des Lehrgebietes Heil- und Sonderpädagogik zur sogenannten Edu-Kinestetik.* 1997. Quoted in Rychetsky, „Kinezjologia edukacyjna."

[361] C. Kant, "Befreite Bahnen, Behinderte Bahnen? Kritische Bemerkungen zur Edu-Kinesthetik," *Berichte zur Sondererziehung und Rehabilitation*, 5, (1997): 7-22, http://userpages.uni-koblenz.de/~proedler/res/ata.pdf

[362] E. Breitenbach, "Wie frei sind die Bahnen? Edu-Kinestetik aus Empirischer Sicht," in *Entwicklungen, Standorte, Perspektiven. Sonderpädagogischer Kongress.* ed. D. Schmetz and P. Wachtel (Würzburg: vds., 1998), 195–203.
[363] Eu Kommision, *Ständiger Interministerieller Ausschuss zur Bekämpfung von Sekten* (MILS, 2001), 20, http://home.snafu.de/tilman/mils2001german.pdf.
[364] S. H. Witcher, *Effects of Educational Kinesiology, Previous Performance, Gender, and Socioeconomic Status on Phonological Awareness Literacy Screening Scores of Kindergarten Students* (Virginia: Faculty of the Virginia Polytechnic Institute and State University, Blacksburg, 2001), http://scholar.lib.vt.edu/theses/available/etd-04212001-144531/unrestricted/BrainGymPaper.pdf
[365] B. Goldacre, "Work out your Mind," *The Guardian*, (12 June, 2003), http://www.theguardian.com/science/2003/jun/12/badscience.science.
[366] Goldacre, "Brain Gym®."
[367] Goldacre, "Exercise the Brain."
[368] U. Goswami, "Neuroscience and Education: from Research to Practice?" *Nature*, 7, (May, 2006): 406–413.
[369] Komitet Neurobiologii Polskiej Akademii Nauk, *Opinia Dotycząca Podstaw Naukowych Metody „Educational Kinesiology" oraz Konsekwencji jej Stosowania* (Warszawa, 20 October, 2006).
[370] UK Economic and Social Research Council's Teaching and Learning Research Programme, *Neuroscience and Education: Issues and Opportunities.* (3, September, 2007), http://www.tlrp.org/pub/documents/Neuroscience procent20Commentary procent20FINAL.pdf
[371] K. J. Hyatt, "Brain Gym® - Building Stronger Brains or Wishful Thinking?" *Remedial and Special Education*, *28*, (2007): 117–124.
[372] Ibid.
[373] Korab, *Kinezjologia Edukacyjna.*
[374] A. Grabowska, "Kinezjologia Edukacyjna w Świetle Najnowszej Wiedzy o Mózgu," in *Kinezjologia Edukacyjna: Nauka, Pseudonauka czy Manipulacja?* ed. K. Korab (Warszawa: Instytut Badań Edukacyjnych, 2008), 42.
[375] M. Borkowska, "Opinia Dotycząca Educational Kinesiology," in *Kinezjologia Edukacyjna: Nauka, Pseudonauka czy Manipulacja?* ed. K. Korab (Warszawa: Instytut Badań Edukacyjnych, 2008), 77.
[376] E. Pisula, "Kontrowersje Wokół Stosowania Educational Kinesiology w Terapii Dzieci z Autyzmem," in *Kinezjologia Edukacyjna: Nauka, Pseudonauka czy Manipulacja?* ed. K. Korab (Warszawa: Instytut Badań Edukacyjnych, 2008), 72.
[377] Z. Kułakowska, "Dysharmonia Rozwojowa. Fakty Neuropediatryczne Wobec Zapewnień Educational Kinesiology," in *Kinezjologia Edukacyjna: Nauka, Pseudonauka czy Manipulacja?* ed. K. Korab (Warszawa: Instytut Badań Edukacyjnych, 2008), 61.
[378] P. G. Dennison, and G. E. Dennison, *Brain Gym: Teacher's Edition* (Ventura, CA: Edu Kinesthetic, 1989).
[379] R. Borowiecka, "Kinezjologia Edukacyjna w Polsce. Wybrane Aspekty Metody i Zasięg Działania Ruchu," in *Kinezjologia Edukacyjna: Nauka, Pseudonauka czy Manipulacja?* ed. K. Korab (Warszawa: Instytut Badań Edukacyjnych, 2008), 13-40.

CHAPTER 13: WARNING! HUGGING CAN BE DANGEROUS TO LIFE AND HEALTH! ATTACHMENT THERAPY

God used intrusive techniques himself.

—Foster Cline

When Alosha, the youngest Karamazov brother in Dostoevsky's eponymous novel, finally manages to seek out his elder brother Ivan, who has been hiding from him, they engage in a long and deep conversation in which Ivan probes the most difficult areas, including the nature of human cruelty and cruelty against children in particular. During their exchange, Ivan Karamazov draws on his rather curious collection of press cuttings, stories, and individual accounts he has amassed over the years. Among them are reports of the most elaborate ways in which rampant soldiers murder children before their mothers' eyes, but also stories where the main roles are played by modern civilized Europeans. Here is one of them:

> There was a little girl of five who was hated by her father and mother, "most worthy and respectable people, of good education and breeding." You see, I must repeat again, it is a peculiar characteristic of many people, this love of torturing children, and children only. To all other types of humanity these torturers behave mildly and benevolently, like cultivated and humane Europeans; but they are very fond of tormenting children, even fond of children themselves in that sense. It's just their defenselessness that tempts the tormentor, just the angelic confidence of the child who has no refuge and no appeal, that sets his vile blood on fire. In every man, of course, a demon lies hidden—the demon of rage, the demon of lustful heat at the screams of the tortured victim, the demon of lawlessness let off the chain, the demon of diseases that follow on vice, gout, kidney disease, and so on.
> This poor child of five was subjected to every possible torture by those cultivated parents. They beat her, thrashed her, kicked her for no reason till her body was one bruise. Then, they went to greater refinements of cruelty—shut her up all night in the cold and frost in a privy, and because she didn't ask to be taken up at night (as though a child of five sleeping its angelic, sound sleep could be trained to wake and ask), they smeared her face and filled her mouth with excrement, and it was her mother, her mother did this. And that mother could sleep, hearing the poor child's groans! Can you understand why a little creature, who can't even understand what's done to her, should beat her little aching heart with her tiny fist in the dark and the cold, and weep her meek unresentful tears to dear, kind God to protect her?[380]

Is this a kind of image that could only be born in the mind of an author famous for his capability to explore the murkiest corners of human nature? Perhaps he overheard the story somewhere? And if so, could this story have only happened in "savage" Russia, or perhaps also elsewhere in Europe? Is it thinkable that a similar scenario could occur today?

It is questions like these that this chapter will explore. In it I will focus on a therapy which makes the methods discussed previously look like the innocent games of a lunatic scoutmaster: holding therapy. The core feature of holding therapy is hugging, and its intended purpose is to build or rebuild an emotional attachment between the mother and the child through close physical contact such as hugging and stroking. The therapy works from the principal assumption that the child's main problem is that it lacks a sense of security in its relationship with the mother. The therapy is often used to correct the behavior of autistic children, adopted children, and children with reactive attachment disorder. But can anything be wrong with hugging?

The answer, sadly, is yes. Holding therapy has demonstrated its damaging and lethal potential in numerous cases. Yet before we examine these consequences, let us first look at the specific features of the therapy:

> It usually has three phases: confrontation, rejection, and resolution.
> Confrontation is a brief phase during which you and your child come together to initiate *holding*. It begins when you take your child on your lap. It may begin as a quiet time of comfort and closeness. If, on the other hand, this is an unplanned session that you initiate because of misbehavior or a tantrum, then it may progress quickly, even instantaneously, to the rejection phase. In the first stage, you and your child are each tuned in to your own thoughts and feelings. As you each move on to try to be in touch with the other, you both begin to feel the hurt and anger provoked either by the barriers you experience or by specific clashes you have had in the previous hours or days. The child may also express hurt or anger over actions by friends, siblings, or teachers. ...
> As you continue to hold, confrontation will either gradually or suddenly lead to rejection, the second stage of *holding time*. Your child will begin to resist you. He may fight you physically, verbally or both. As you hold on despite the child's entreaties to let him go, his full range of feelings will begin to emerge, usually starting with anger but sometimes with fear or sadness. There follow one or more cycles of anguished outbursts alternating with quieter moments during which the child may be calmer and rejection seems to wane. A successful *holding* will not be complete, however, until the calm period achieves true resolution, the final phase.
> Resolution is marked by a sweet molding together of the mother and child, physically and emotionally. The pair relax in each other's arms, gaze into each other's eyes, and offer tender physical caresses.[381]

The second of the three phases is crucial for our discussion and we will now focus on what actually happens during that time. Most children subjected to this therapy will offer stiff resistance to avoid either physical or eye contact. They will attempt to break free, cry, scream; possibly kick, scratch, and spit. Some children vomit and urinate themselves. The mother is instructed to demonstrate her physical advantage and use force. She is encouraged by the therapist also to vent her emotions; she may cry and scream just as hard as the child does. It is believed that this kind of demonstration of the mother's physical superiority will make the child feel more secure, and as this phase progresses towards its end the child will gradually become quieter and calm down. This short account barely touches on the finer details, but it already reveals a picture that may be less alluring than a layperson's first idea about how hugging can help out children and parents in need.

Holding therapy belongs to a group collectively known as attachment therapies. They share similar theoretical premises and include also rebirthing,[382] anger-reduction therapy and the Evergreen model, compression therapy, corrective attachment therapy, and many others. It is beyond the scope of this modest chapter to define these therapies or to discern between them; nor is it essential here to provide an exhaustive list of them. The point is that the common objective of all of them is to correct attachment in children, and they share very similar premises. From now on, whenever I mention attachment therapy I shall mean either a specific method or several of them used in conjunction, a favorite approach among attachment therapists.

Supporters of attachment therapy work from the premise that emotional attachment between parents and children originates as early as the pregnancy stage, and then develops after birth through cycles involving frustration and subsequent gratification. Hunger, cold, wetness, unpleasantness; these are the frustrations. When the parents feed and care for the child, frustrations are replaced by gratifications. These gratifications are strongly linked with the eye contact that parents maintain with the child during these activities. The founders of this therapy claim that rejection during the prenatal period or depriving the child of gratifications after birth are the principal causes of an emotional barrier to attachment that accumulates as unrealized rage.[383] However, repressed and blocked rage may also manifest itself through other symptoms than mere inability of attachment. These symptoms include: disobedience, failure at school, late development of language skills, so-called mirror writing, and many others.[384]

Therapists have targeted this repressed anger, which they claim must be released in a cleansing catharsis. To ensure the desired effect they cause pain and a strong sense of fear in the child. They believe that if they can provoke a fit of rage the child will pass through a process of purification and will become capable of experiencing positive feelings, such as joy, gratitude, and obedience. But if the little patient resists the therapy, complains, cries, coughs, or vomits, these behaviors are interpreted as symptoms of resistance. In these cases attachment therapists recommend maintaining or increasing the child's discomfort to the point of a catharsis.[385] They also suggest that fits of anger should be regarded as the child initiating a "power struggle" against the mother or father. In their opinion, the parent simply must not lose this struggle. This approach is best illustrated by a quote from Foster Cline, a paramount figure among attachment therapists: "here is no compromise – it's either win completely or lose absolutely."[386]

Recommendations and guidelines on ways to trigger a "purifying" fit of anger and how to conduct these skirmishes with the child reach beyond the clinic and extend into the home. We shall now focus on a number of cases reported in the United States to learn just what kind of practice can develop from an attachment therapist's inspiration.

Theresa and Reed Hansen applied attachment therapy recommendations to their two adopted children, a boy of four and a girl of five. The parents tried to elicit obedience by locking both children up in the bathroom. For several hours at a time, sometimes overnight. No bed sheets, no blankets. Yet it was food control, a popular method among parents operating on recommendations of attachment therapists, that was their tool of choice to help win the "power" skirmishes. Misbehaving was punished by food deprivation for up to several days at a time. This was intended to teach them to obey. The "therapeutic" objective was never achieved, as the starving children, in a state close to death, were taken away and the parents hauled into court.[387]

Joseph and Yevette Heiser, father and stepmother, used similar methods on their daughter. The child had to earn her meals by standing outside holding 3.5 pound weights with her arms extended in front of her for ten minutes, or twenty minutes if she wanted dinner. The seven-year-old turned to begging for food at school and when she was found rummaging through a trash can for scraps of food she weighed only 38 pounds and was diagnosed with severe damage to the liver due to starvation. Her health condition was not caused by poverty, as the family was leading a relatively comfortable life, but by the implementation of attachment therapists' recommendations.[388]

Food control is not the only method used by parents convinced that their children suffer from an attachment disorder and believe that attachment therapy is the only way out. Sharen and Michael Gravelle, inspired by advice from attachment therapist Elaine Thompson, kept ten of their eleven adopted children in cramped wooden cages connected with an alarm system that activated whenever they wanted to leave. The urine-stinking cages also served as night quarters, where they would often have to get by without bed sheets. Cages were also used as a form of punishment for disobedience. Other means used by the adoptive parents included food control, as they locked up their fridge and pantry, forcing children to copy the entire Bible by hand, and watering the garden in the winter. One of the children had its head thrust into the toilet, while the night wetting problem of another child meant it was forced to live in the bathroom and sleep in the bathtub for 81 days. Another child's gaping mouth, which the parents associated with Down syndrome, ended up having a sock thrust in it. In return for fulfilling their duties as adoptive parents the Gravelles received free health care and other social benefits, as well as around 100 thousand dollars per year. When they were finally brought before a court, the oldest of the adopted children asked for the parents to be locked up for the same period as all the children had to spend in the cages.[389]

Yet if these children went through a horrific ordeal, others were even less lucky. Ten-year-old Candace Newmaker was first subjected to holding therapy and then to rebirthing, another attachment therapy. Rebirthing involves complete physical restraint, from which the patient is to free herself in order to experience a symbolic rebirth. Again, the treatment is intended to free repressed emotions. To achieve this effect, Candace was wrapped in sheets and laid down on the floor by assisting adults, who then covered her with cushions and leaned on them. She was told to free herself and be reborn. For 40 minutes, the girl begged for air, vomited and choked, which the therapists interpreted as active resistance. For a subsequent 30 minutes, she remained silent and motionless. When she was finally removed from under the cushions and unwrapped, she was already blue and attempts at resuscitation failed. She had choked to death on her own vomit. The girl's stepmother was present throughout the "treatment."[390] The whole session was video recorded, thus providing a detailed account which was shown during the court trial.

The Candace Newmaker case remains the best-known case of death as a result of the application of attachment therapy. This is probably because it occurred in the presence of several adults, including the stepmother, was recorded, and played back many times. Sadly, the grim list of the therapy's death toll does not end there. Here is its continuation:

- Andrea Swenson, died in 1990 at the age of 13.
- Lucas Ciambrone, died in 1995 at the age of 7.
- David Polreis, died in 1996 at the age of 2.
- Krystal Tibbets, died in 1997 at the age of 3.
- Roberta Evers, died in 1998 at the age of 7.
- Candace Newmaker, died in 2000 at the age of 10.
- Viktor Matthey, died in 2000 at the age of 7.
- Logan Marr, died in 2001 at the age of 5.
- Cassandra Killpack, died in 2002 at the age of 4.

There are also several cases of children saved from the same fate at the very last moment:

- Jeannie Warren, saved in 1990.
- Carol Abbott, saved in 1994.
- Bruce, Keith, Tyrone, and Michael Jackson, saved in 2003.
- Four Vasquez siblings, saved in 2007.[391]

All of these children, battered to death or barely saved, had been subjected to at least one kind of attachment therapy. There is no certainty that these two lists are exhaustive, as they only contain well-known cases reported in the press and widely commented in the media. But they inspire a number of important questions. Were these just "workplace incidents"? Maybe they were a consequence of overzealous therapists and parents applying their guidelines? Perhaps there is an error inherent in attachment therapy which leads to such tragic outcomes? In the section below I will examine these questions.

A close analysis of attachment therapy reveals two mechanisms embedded in it, most probably inadvertently, but possibly as an emanation of the founders' own needs, that are otherwise known to lead people to bestiality: *obedience to authority* and *desensitizing*. The reader might be familiar with the famous experiment performed by Stanley Milgram that illustrated obedience to authority.[392] In the experiment, normal people found themselves capable of applying even lethal doses of electric current to innocent victims simply because they were told to do so by a person they had accepted to be an authority. The participants crossed these boundaries even though every one of them had declared beforehand that they would never do it.

Let us now go back to holding therapy. The mother or guardian presses the child to her or his chest, but the child is protesting. A normal reaction would be to let it go, but the therapist, speaking in a calm voice not unlike one of Milgram experimenters', says: "please continue." The child begins to cry, cough, vomit, but the therapist intervenes: "you can't interrupt now, you must continue." When things turn even uglier and the child wets itself and chokes, the parent hears something akin to the previously quoted words of Foster Cline: "it's either win completely or lose absolutely." In Milgram's experiment, all that was needed for a huge majority of participants to subordinate to the instructions of the experimenter was a sign featuring the name of a university, a white lab coat, and vestiges of science in the procedure. In therapy there are many more attributes that strengthen this effect. It is the therapist's experience, his or her diplomas and certificates on the wall, the *knowledge* that this is an actual specialist, someone who has written books, and, last but not least, the fact that the parents have chosen this particular professional on a recommendation from someone else. Where, as often happens, the therapy is intended for an adopted child yet another factor conspires to help the adoptive parent yield to the authority and, consequently, to allow even greater harm: the strength of the emotional bond with the child. Whatever we may think about the intentions and feelings of adoptive parents, this bond is likely to be weaker than with a parent's own biological child that he or she has raised from birth. Indeed, when adoptive parents knock on an attachment therapist's door, this child will have most probably already got up their noses, or they would not have been there in the first place.

Attachment therapy, and holding therapy in particular, also perform a desensitizing function. This concept has nothing to do with the medical term reserved for allergy treatments and has only a little to do with the similar term "desensitization," which denotes a behavioral therapy technique. Desensitizing is a procedure used in training soldiers, intelligence agents, and hit men, and in other professions where the presence of other people's suffering and/or causing this suffering is a highly likely scenario. Desensitizing a soldier to human suffering involves bringing him or her to a state where witnessing mortal agony, despair, or inner organs being torn apart makes little or no impression at all. This is achieved, for example, by showing brutal films and images, and by an intentional use of terminology such as "eliminate object" instead of "kill," "expectant" instead of a "dying person," etc.

A rather stunning description of desensitizing is found in Peter Watson's book *War on the Mind: The Military Uses and Abuses of Psychology*, where the author quotes an account of Dr. Thomas Narut, a psychiatrist and a Navy commodore working for the U.S. federal government. He claims to have taken part in the development of a desensitizing technique that involved soldiers being shown increasingly graphic scenes of brutality. Narut also claims that during the training soldiers' heads were kept in a brace that prevented them moving and a special device was used to prevent the soldiers' eyelids from closing[393]. The U.S. authorities denied using these procedures on many occasions, but reports of similar "training" come from too many sources and locations worldwide to doubt that such desensitizing techniques are in active use. Others report on the development of chemicals intended to reduce soldiers' natural resistance to killing.[394] A combination of desensitizing training and potential chemical support would quickly and effectively steer natural human defense mechanisms towards indifference to human suffering.[395]

The similarity between this procedure and attachment therapy is startling. Even if the founders of the method never dreamed of producing this effect, and there is no reason to think otherwise, the actual outcomes of the therapy discussed above have shown that these mechanisms are indeed at work.

While obedience to authority and desensitizing are probably inadvertent consequences, the numerous references of attachment therapy to psychoanalysis are clearly deliberate and intentional. They include, for example, repression of anger, the need to free emotions through catharsis, and the hydraulic concept of emotions, according to which what has been locked up and compressed must come out under its own pressure sooner or later. Jean Mercer, one of the top experts on and critics of attachment therapy, has tracked these theoretical inspirations to Wilhelm Reich, one of the more colorful figures of psychoanalysis, and specifically his book *Character Analysis*, in which Reich expressly wrote about the significance of eye contact and the need to physically intervene with the patient.[396] Incidentally, the quoted book seems to be his only work treated seriously by his fellow psychiatrists.[397]

Other sources of theoretical inspiration for attachment therapies include, according to Mercer, late transactional analysis[398] and the New Age movement.[399] It is widely accepted that the way to the development of attachment therapy was paved by Robert Zaslow and his "Z-process," which he developed in the 1970s.[400] Zaslow tried to elicit attachment in autistic children by triggering their rage through hugging against their will. He believed that this would break through their defenses and make them more open to others. He thought that attachment would emerge when the child, having experienced pain, fear, and anger, is freed from these feelings upon making eye contact with its guardian. Without experiencing a sequence of events that involves being freed from fear and anger the child would not be capable of attachment and maintaining eye contact with others. Holding therapy is a straight-line descendant from Zaslow's practices. The "holding" component here is not intended to provide a sense of security, but to elicit negative emotions and trigger the said sequence of events. Zaslow retired from his career as a psychologist and lost his therapist license after being prosecuted in a case for causing bodily harm to a patient during rage-reduction therapy. The demise of Zaslow and the subsequent discovery of biological causes of autism spelled the end of the use of the Z-process and holding therapy in autism treatment.[401] Or so it seemed.

Meanwhile, the charming little town of Evergreen, Colorado, at the foot of the Rocky Mountains, saw the opening of a private clinic by Foster Cline, a therapist heavily influenced by Robert Zaslow. Cline adapted Zaslow's key notion, derived from psychoanalysis, about "breaking through" a child ego's defenses. Soon, "breaking through" would become the dominant metaphor used by therapists working with children with a weakened sense of attachment. Cline's clinic, known as the Attachment Center, advocated the use of methods similar to those used in holding therapy for the treatment of adopted and maltreated children. The approach began to take root and similar clinics sprang up like mushrooms. The Evergreen-based clinic trained numerous therapists, including Connell Watkins, who would rise to become a partner and head of the Attachment Center and, in 2001, would face charges in the Candace Newmaker case, alongside other therapists. Foster Cline, regarded as the "founding father of attachment therapy," voluntarily gave up his therapist license and moved to Idaho as a consequence of an investigation into an incident that had occurred during one of his therapies.

The attachment therapy movement gained a hugely influential ally in Niko Tinbergen, the Dutch ethologist and Nobel laureate, who published a book on autism in 1983.[402] Based on his own observations of bird behavior and on the observations of Konrad Lorenz, another famous Nobel Prize laureate, Tinbergen concluded that autism had to be a result of an error in the formation of the original bond between the mother and child. He encouraged the use of holding therapy even though he had no scientific basis for doing so. This is an example of a dark chapter in the history of ethology, whose most eminent representatives contributed to the consolidation of this and several other superstitions in the understanding of human behavior. Tinbergen and Lorenz left yet another "legacy" in the notion of the "hydraulic nature of emotions." Originally proposed by psychoanalysts and then supported by the authority of such eminent ethologists, this concept has somehow managed to linger on. In attachment therapy it has been employed in the utterly false idea mentioned above that accumulated anger must at all costs be released.

Tinbergen's book on autism and his endorsement for the attachment therapy movement led to the success of Martha Welch, a child psychiatrist who recommended holding therapy in autistic children. In her book *Holding Time*,[403] Welch claims that autism is a result of failures in the attachment relationship with the mother. Tinbergen contributed a gushing foreword to her book and the name of the celebrated Nobel laureate helped pave the author's road to fame. Supporters of attachment therapy also claim that their approach grew out of John Bowlby's well-established attachment theory. In his classic trilogy, one of the most frequently-quoted studies of contemporary psychology, Bowlby expressly rejects the classic Pavlovian and psychoanalytical models, where attachment is depicted as an effect of feeding. He asserts that infants seek closeness, touch, and a sense of security from their caregivers. The development of attachment depends on the caregiver's responsiveness, i.e. his or her sensitivity and the manner of their reactions to the child's actions. According to Bowlby, the attachment thus developed is also significant for the person's subsequent contacts with others, in relationships characterized by friendship, love, etc.[404] However, holding and similar therapies have nothing to do with Bowlby's attachment theory. Followers of holding therapy are trying to appropriate the terminology and use purely formal measures to give the impression that they are representing a practice based on reliable research. The actual theoretical foundations of attachment therapy are nothing but beliefs and subjective convictions which lack any empirical basis.

The references to Bowlby's valuable concept made by attachment therapy writers with increasing frequency[405] are but desperate attempts at a *post hoc* legitimization of an extremely controversial therapy that finds itself irreversibly sliding towards the threshold of societal and scientific acceptability. Neither a robust theoretical basis, nor results of research (despite what can be read on the supporters' websites) provide an effective anchor that would prevent holding therapy from universal disapproval and oblivion. Indeed, all it is resting on are the recommendations of a handful of parents and therapists.

Any attempt at investigating attachment theory's actual efficacy runs into obstacles as early as the phase of seeking reliable measures. For example, data collected by attachment therapy supporters come only from caregivers' reports. Therefore a statement such as "the child has eye-contact problems" may in fact mean that "the child does not want to make eye contact when the caregiver demands it." Surveys used by therapists to investigate the attachment disorder syndrome suffer from similar design problems. They only involve subjective observations by parents and are not verified by objective observers. It is no wonder then that the parents like the diagnosis – they are in fact its authors! Could this be one reason why attachment therapy centers have enjoyed such great popularity?

Deficiencies in the design of survey questionnaires are not the only problem found in current research into attachment therapy efficacy. Most of the measurements of attachment therapy effectiveness performed at centers using this method were executed by persons directly involved in the therapy. None of these measurements met the requirements listed in Chapter 10 of this book that were developed by a group of experts known as the Task Force on Promotion and Dissemination of Psychological Interventions appointed by the American Psychological Association.[406]

The previously mentioned Jean Mercer conducted a thorough inquiry into the subject as she searched for studies that would verify the efficacy of attachment therapy.[407] Here is what she found. The two doctoral theses cited by authors writing on attachment therapy were nowhere to be found. They did not even feature in the register *Dissertation Abstracts International* that publishes abstracts of doctoral theses. Eventually, when Mercer stumbled upon one of them, it turned out to be about a completely different topic. Despite countless anecdotal accounts of attachment therapy's efficacy she encountered not a single reliable case study that would support them. The sole one that she found in Cline's book was not about attachment therapy. Mercer identified just three quasi-experimental research studies, only one of which was performed by a researcher not linked to attachment therapy centers but did not include any statistical analysis. All three, however, suffered from methodological errors, and where statistical analyses had been performed Mercer found them to be of debatable quality to say the least.

In the face of the scale of potential damage, its lack of sound theoretical foundations, lack of reliable diagnostic tools, or evidence of its efficacy, many organizations and individuals publically declared an expressly negative position towards attachment therapy. This turning of the tide occurred only after reports on the first fatalities, but the real breakthrough came about after publication of a book by psychologists Jean Mercer, Larry Sarner, and Linda Rosa titled *Attachment Therapy on Trial: The Torture and Death of Candace Newmaker*.[408] The three authors had performed the role of court experts in the case for reckless child abuse resulting in Candace Newmaker's death and were so shocked by what they saw that they devoted themselves to fighting against attachment therapy. Apart from the book they also inspired the establishment of Advocates for Children in Therapy and actively contributed to its work.[409] The organization was the first to publish, in 2002, a highly critical position on attachment therapy and is still operating in education, continues to publish its own material, and consistently opposes the application of this therapy.

In 2003, the scientific journal *Attachment and Human Development* released a special issue entirely devoted to attachment therapy. The magazine was a collection of critical papers by established researchers and experts in the area of attachment building. This was followed in 2006 by another important and exhaustive publication, this time a report by the American Professional

Society on the Abuse of Children, an organization that had made it their mission to counter abuse of children[410]. In 2007, Scott Lilienfeld, in an article published in *Perspective on Psychological Science*, included holding therapy on a list of the most threatening and dangerous therapies to patients.[411]

Two American states, Colorado and North Carolina, officially banned rebirthing,[412] while Utah banned holding therapy.[413] A list of organizations opposing to the application of attachment therapy and methods of intervention in children propagated by its supporters is very long. It includes the U.S. Supreme Court, the Congress, President Bill Clinton, the American Psychiatric Association, the American Psychological Association, and dozens of other organizations and bodies. A full list of American and non-American organizations protesting against the use of attachment therapy would cover several pages of this book. Adding the names of well-established psychologists, physicians, scientists and journalists that also oppose the kind of treatment of children that takes place in attachment therapy would swell my book considerably. Is this not enough to put attachment therapy on a list of forbidden therapies?

The answer is "no." Attachment therapy is too good and too simple a business for it to disappear so easily. Both the centers that use attachment therapy and the therapists themselves are doing very well indeed (perhaps with the exception of those few currently serving prison sentences). After a few court processes and publications they may have been temporarily thrown on the defensive, but they are clearly not ready to give up just yet. Their new Association for Treatment and Training in the Attachment of Children (ATTACh) organizes conferences, carries out educational work, and publishes self-help books. Its website offers dozens of addresses of attachment therapists and clinics in the United States,[414] and these are just those represented by ATTACh members. It would be difficult to estimate the number of non-associated therapists in this field.

Attachment therapy is not only used in the USA; many British therapists use these methods too.[415] Elsewhere in Europe it would be difficult to estimate the scale of such practice, but it is certainly present, for example, among Polish therapists. The service dobrodziecka.pl ("the good of the child") offers this encouraging message:

> The therapy centers operate on the basis of the most successful therapeutic methods, including:
> - The holding therapy method that involves coercing children into wanting to maintain physical contact with caregivers;
> - Attachment therapy, or observation of the child's behavior and joining in without criticism, assessing, instructions or banning; this makes the child get used to a new person and new situation.[416]

The educational web portal edux.pl also encourages holding therapy. As it postulates: "Coercive holding therapy must be practiced regularly. It would be difficult to say for how long. Indeed, the desired result depends on many factors, including the scale of the disorder, consistency of the mother and competence of the therapist. Typically, this takes years. Often, mothers find themselves unable to bear the emotional tension and quit. Holding is a safe method and there are no counter-indications for its application."[417]

As I read these descriptions I begin to doubt whether what I have written above has any bearing on reality. Whether these starved children scavenging among trash cans, sleeping in cages, locked up in bathrooms, or choking on their own vomit are but figments of my imagination and its cruel tricks. Perhaps I was writing about something completely different than what is being described so convincingly on these websites? I check again and make sure; no, the names of the founders of these therapies are the same, the places where the ideas grew from are the same, the terminology is the same, and so is the therapy...

Why, in the face of such crimes perpetrated by attachment therapists and by parents who scrupulously follow their recommendations, do so many people and organizations not just fail to reject these therapies with repulsion as incompatible with their basic values, but brag about them in public? Without a sense of shame or fear of consequences?

Let us consider a range of possible answers to these questions. First, these people must be unprincipled and care mostly about the business outcome of the therapy. Second, these people know nothing about the victims and potential adverse consequences of the therapy and are generally ignorant. Third, they realize what has happened, read the criticism, but have devoted so much of their careers to attachment therapy that doubting what they have been doing would cause them tremendous cognitive dissonance. Therefore, they clutch to their beliefs and cognitively select incoming information.

I reject the first answer as the least likely. There are many other professions where a lack of scruples offers far more benefits for such people to end up in a therapeutic practice. And even if they do, they are few and far between. These kinds of motivations may at most have driven the founders of some pseudo-therapeutic methods to build business empires around them. The third answer seems highly likely with regards to people who have been "stuck" in these therapies for years and it is close to certain in relation to therapy founders and its leading proponents. At work here are mechanisms so aptly described by Carol Tavris and Elliot Aronson in their *Mistakes Were Made (But Not by Me)*. True, the book is not about the founders of attachment therapy, but about proponents of the recovered-memory syndrome therapy, which is equally if not more destructive. Still, the same mechanisms are at play in both cases: "Today, standing at the bottom of the self-justifying pyramid, miles away professionally from their scientific colleagues, having devoted two decades to promoting a form of therapy that Richard McNally calls 'the worst catastrophe to befall the mental-health field since the lobotomy era,' most recovered-memory clinicians remain as committed as ever to their beliefs. How have they reduced their dissonance?"[418]

The second option seems to me the most likely one. Ignorance is known to be the driving force behind most pseudoscientific disciplines in general and it has the potential to power any movement, no matter how absurd. Evidence in support of this option comes from reading relevant websites. Most of them carry the exact same description, sometimes also repeating the very same errors. Clearly, their creators and administrators copy and paste the content from somewhere else. Their belief in authority is so strong that it is sufficient for them to have heard about holding therapy during their university years to accept that the method is universally established. Can ignorance serve to vindicate them? Certainly not, as nothing vindicates people who may be harming others. They are certainly not vindicated by their good will or a desire to help others. Those who lack the intellectual potential or motivation to properly learn their selected field of knowledge should never become therapists, let alone children's therapists. If bringing help to others is their chosen vocation, they would do a better job by assisting children across the street or working in a school canteen than by thrusting them into the grip of holding therapy. Finally, not even the halo surrounding those who bring help to the needy makes them immune to justice, as many therapists in the USA have learned the hard way.

Attachment therapy is based on a lie. The phenomena and processes discussed by its supporters either do not take place at all or take a different course than that described by them. There is no available scientific research that can confirm the efficacy of these therapies. Their popularity was only possible thanks to a clever trick that took responsibility off the parent's shoulders, especially adoptive parents. Indeed, a parent who has problems with an adopted child is told that he or she is not to blame. The source of the problem lies in the child and its repressed anger. The therapist encourages not just the child, but also the parent to release that anger.

Finally, therapists like to quote yet another cause of attachment disorders, fetal alcohol syndrome. A pregnant mother drinks alcohol, thus poisoning the fetus which, according to attachment therapists, leads to problems with rearing. This justification was often adopted by the defense in court cases brought against parents, especially adoptive parents, and/or therapists. This would have been partially excusable if an actual fetus alcohol syndrome had been identified, but this explanation was systematically abused to explain problems with obedience. The secret of the popularity of this line of defense lies in the shifting of the responsibility from the adopted parents onto the biological mother. And if the mother was not drinking, there was always an array of other factors that might have led to problems. Indeed, the very thought of abortion in the pregnant mother's mind could, at least according to supporters of holding therapy, prevent normal development of attachment. They really claim that the fetus receives not just the mother's emotions but also her thoughts, and if they are negative the child's emotions will be "stifled."

What a brilliant concept! The parents assess the child's behavior, their assessment is then used for diagnosis, causes of problems are always located outside of the parent, negative emotions are always justified, while the pain and misery of the child are just necessary therapeutic measures. How different this approach is from that involved in Bowlby's attachment theory, where it is the caregiver, and specifically his or her responsiveness, that determines the development of attachment. Attachment-based therapy, a name deliberately chosen to differentiate this method from attachment therapy, provides an alternative to the methods described above. It may not be free of its own problems, but at least its methods are derived from the far-better documented Bowlby theory, are researched, and constantly improved.[419] If nothing else, it doesn't result in children being tortured or murdered by well-intentioned but misguided parents, which is enough to make it infinitely more commendable than attachment therapy.

---

[380] F. Dostoyevsky, *The Brothers Karamazov* (12 February, 2009), 265, http://www.gutenberg.org/ebooks/28054

[381] M. G. Welch, *Holding Time: How to Eliminate Conflict, Temper Tantrums, and Sibling Rivalry and Raise Happy, Loving, Successful Children* (New York: Simon and Schuster, 1989), 46-47.

[382] This name is far more frequently associated with a therapy developed by Leonard Orr that employs yoga breathing techniques. In its form used by attachment therapists, rebirthing has little to do with the former therapy and is a far more dangerous activity, as will become apparent in this section.

[383] F. W. Cline, *Hope for High Risk and Rage Filled Children* (Evergreen: EC Publications, 1992).

[384] Welch, *Holding Time*.

[385] K. Reber, "Children at Risk for Reactive Attachment Disorder: Assessment, Diagnosis, and Treatment," *Progress: Family Systems Research and Therapy 5*, (1996): 83-98.

[386] F. W. Cline, *Conscienceless Acts Societal Mayhem: Uncontrollable, Unreachable Youth and Today's Desensitized World* (Golden. CO:The Love & Logic Press 1995), 177.

[387] http://www.childrenintherapy.org/victims/hansen.html

[388] http://www.childrenintherapy.org/victims/heiser.html

[389] http://www.childrenintherapy.org/victims/gravelle.html

[390] J. Mercer, L. Sarner, and L. Rosa, *Attachment Therapy on Trial: The Torture and Death of Candace Newmaker* (Westport: Praeger Publishers, 2003).

[391] http://www.childrenintherapy.org/victims/

[392] S. Milgram, "Behavioral Study of Obedience," *Journal of Abnormal and Social Psychology 67*, (1963): 371-378.

[393] P. Watson, *War on the Mind: The military Uses and Abuses of Psychology* (London: Hutchinson, 1978).

[394] E. Baard, "The Guilt-Free Soldier: New Science Raises the Specter of a World Without Regret," *Village Voice*, (22, January, 2003).
[395] Watson, *War on the Mind.*
[396] J. Mercer, "Attachment Therapy. A Treatment Without Empirical Support," *The Scientific Review of Mental Health Practice 1*, (2002): 9-16.
[397] Wilhelm Reich, born near Tarnopol (present day Ukraine) in the Austrian partition of Poland, met Sigmunt Freud at university and soon became his protégée, rising in the ranks of the Vienna Psychoanalytical Society. He was also active in the Austrian Socialist Party. There he was criticized for his concept of a life force derived from orgasms and broke away from socialists, joining the communists instead. As a Freudian-Marxist he claimed that it was proletariat's sexual sterility that was to blame for the disappearance of its political awareness. Sadly, he found no followers among communists either. His work was labeled as "anti-Marxist rubbish" and as a result Reich split from them as well. Differences of opinion between Reich and Freud and his followers lead to Reich's expulsion from the International Psychoanalytic Society in 1934. Following several more break-ups and splits, the constantly criticized Reich moved to the USA, where he established his own Orgone Institute, where he worked on his orgone energy concept. Charged with conducting illegal research, Reich died in prison in 1957. Years after his death, he would become one of the counter-cultures' ideologists. For more see: M. Gardner, *Fads and Fallacies in the Name of Science* (New York: Dover Publications, 1957)
[398] J. L. Schiff, and B. Day, *All my Children* (New York: Jove Pubns, 1979). As cited in Mercer et al., *Attachment Therapy*. Transactional analysis is a concept of human relationships proposed by Eric Berne based on the assumption that every person's ego could be in one of three states: child-like, adult-like, and parent-like.
[399] W. R. Emerson, "Points of View: The Vulnerable Prenate," *Pre- and Perinatal Psychology Journal 10*, (1996) 125-142.
[400] R. Zaslow, and M. Menta, *The Psychology of the Z-Process: Attachment and Activity* (San Jose: San Jose University Press, 1975).
[401] M. L. Speltz, "Description, History and Critique of Corrective Attachment Therapy," *The APSAC Advisor 14*, (2002): 4–8.
[402] N. Tinbergen, and E.A. Tinbergen, *Autistic Children: New Hope for a Cure* (London: Allen & Unwin, 1983).
[403] Welch, *Holding time.*
[404] J. Bowlby, *Attachment and Loss, vol. 1, Attachment* (New York: Basic Books, 1969/1982); J. Bowlby, *Attachment and Loss, vol. 2, Separation: Anxiety and Anger* (New York: Basic Books, 1973); J. Bowlby, *Attachment and Loss, vol. 3, Sadness and Depression* (New York: Basic Books, 1980).
[405] T.M. Levy, and M. Orlans, "Attachment Disorder as an Antecedent to Violence and Antisocial Patterns in Children," in *Handbook of Attachment Interventions*, ed. T. Levy (San Diego: Academic Press, 2000).
[406] D. L. Chambless, M. J. Baker, D. H. Baucom, L. E. Beutler, K. S. Calhoun, A. Daiuto et al., "Update on Empirically Validated Therapies, II," *Clinical Psychologist 51*, (1998): 3–16.
[407] Mercer et al., *Attachment Therapy.*
[408] Ibid.
[409] http://www.childrenintherapy.org/

[410] M. Chaffin, R. Hanson, B.E. Saunders et. al., "Report of the APSAC Task Force on Attachment Therapy, Reactive Attachment Disorder, and Attachment Problems," *Child Maltreatment 11*, (2006): 76–89.
[411] S. O. Lilienfeld, "Psychological Treatments that Cause Harm," *Perspectives on Psychological Science 2*, (2007): 53–70.
[412] Advocates for Children in Therapy, "North Carolina Bans 'Rebirthing,'" *AT News*, (August 2, 2003), http://www.childrenintherapy.org/atnews/2003Aug2.html
[413] Editorial, "Senate, ban Holding Therapy," *Desert News*, (5 February, 2003), http://www.deseretnews.com/article/963086/Senate-ban-holding-therapy.html?pg=all
[414] https://attach.org/attach-resources/registered-clinicians/
[415] J. Sudbery, S. M. Shardlow, and A. E. Huntington, "To Have and to Hold: Questions About a Therapeutic Service for Children," *British Journal of Social Work 40*, (2010): 1534–1552; A. Chaika, *Invisible England; The Testimony of David Hanson* (CreateSpace Independent Publishing Platform, 2012).
[416] http://www.dobrodziecka.pl/informacje/index.php/zdrowie-dzieci/214-orodki-autyzm
[417] http://www.edukacja.edux.pl/p-133-metody-terapii-autyzmu.php
[418] C. Tavris, and E. Aronson, *Mistakes Were Made (But Not By Me): Why We Justify Foolish Beliefs, Bad Decisions, and Hurtful Acts* (Orlando, FL: Harcourt, E. 2007), 123.
[419] L. J. Berlin, C. H. Zeanah, and A. F. Lieberman, "Prevention and Intervention Programs for Supporting Early Attachment Security," in *Handbook of Attachment: Theory, Research and Clinical Applications*, ed. J. Cassidy and P.R. Shaver (New York, London: Guilford Press, 2008), 749–750.

Chapter 14: Dolphins: Wonder Therapists, Intelligent Pets, or Aggressive Predators?

> British swimmer Adam Walker had enough to worry about as he took on the freezing ocean in a grueling eight-hour-and-36-minute swim across the Cook Strait on Tuesday.
> But he had his mind taken off the extreme temperature when he noticed a two-meter shark-shaped figure swimming beneath him in the New Zealand waters.
> Luckily enough, the shark never attempted an attack as a pod of dolphins soon came to his side, swimming alongside Mr Walker for an hour while he crossed the strait.
> It was a magical sight for those watching, as Mr Walker's long strokes were matched by the jumping figures of approximately 10 dolphins, who came so close to the swimmer that he brushed a tail as he swam.

Have we not all read several such accounts where dolphins came to the rescue of humans? Have we all not heard their intriguing love songs? The intelligence of these creatures is so awe-inspiring that we tend to feel a vague sense of kinship and an almost spiritual bond in their presence. We regard the killing of a dolphin as being on a par with the killing of a pet. This sentiment goes back to ancient Rome, where killing a dolphin, believed to be an envoy of the gods, was punished by death.

All of which is perhaps why some of us find it so alluring to believe that these creatures might be better equipped to get in touch with mentally disabled children than us adults. Perhaps it would not be all that difficult to persuade us that dolphins can help these children swim through an ocean of isolation and miscommunication, arriving at the shores of compassion. What if we were also told that the ultrasounds emitted by dolphins go right through our bodies, spreading a healing effect? The website of the Onmega Dolphin Therapy organization has this to say about the therapeutic properties of the dolphins:

> Under the supervision of the doctors and a dolphin trainer the dolphins understand that they should make contact with the patient. This opens possibilities that no medication or doctor could provide. It is most probable that the dolphins can feel the vibrations of a malfunctioning organ through their probing with ultrasonic waves (is being researched). Moreover it has been proved that through the sonic waves the spine and brain come into a resonance that stimulates the production of substances in the nervous system that in turn improve the functioning of the nervous system as a whole. Stress, fear and tension are lessened through contact with a dolphin. One receives positive energy and is freed from negative emotions. ...
> Former therapies cannot nearly match the prodigious improvements in health that are induced by playing with the dolphins. Children suddenly can laugh or move freely that they have not been able to do before. They can speak their first

words, awake laughing out of their half coma or they can do things that they failed to accomplish in previous therapies.[420]

Is this not just wonderful? If you still have doubts, just one look at a list of ailments treated in a dolphinaria should convince even the staunchest skeptic:

> We find that our "Harmony" program can help a wide variety of children: most spectra of autism, manic depression, Rett Syndrome, Tourette's Syndrome, ADHD, Down, hyperactivity.
> We can also help children with more general mental and emotional disorders or difficulties in social adaptation and even children with terminal illnesses (cystic fibrosis, cancer, leukemia). ...
> This program caters to children with serious developmental delays, physical and motor disabilities like Cerebral Palsy, Head and Spinal Cord Injury, Mitochondrial Disorder. Spina Bifida, Muscular Dystrophy, Angelman's Syndrome and other disabilities, which require body work and specialized interventions.[421]

Perhaps we need more of these kinds of centers rather than hospitals, clinics, and other backward health care outlets, which have been finding it so hard to overcome most of these sicknesses?

Dolphins are marine predators that feed mostly on fish and invertebrates. The exception is the largest of the Dolphinidae, the killer whale, which also hunts for seals and penguins. Together with whales, dolphins form the infraorder Cetacea. Indeed, they are incredibly intelligent animals with an encephalization ratio, which is indirect evidence of intelligence, much higher than that of the chimpanzee. Thus equipped, they learn quickly and have been used by humans for entertainment and work. The U.S. Navy has used dolphins as sea-mine detectors, kamikaze fighters, or against enemy scuba divers. Dolphins have yet another feature that makes them rather unique. They use echolocation, or sending and receiving ultrasounds, to find their way in the water.

Dolphins form elaborate social systems with complex group relationship patterns. They are inquisitive and playful. Indeed, many dolphin experts agree that the legendary cases of dolphins saving humans can be explained by the fun-loving nature of these mammals rather than by "intentional" Samaritan gestures. These researchers believe that dolphins may have brought as many shipwreck survivors safely to the beach as they tugged out and abandoned on the high seas. What happened to the swimmer in the story above happens to every sailboat that navigates through waters inhabited by dolphins. These animals will accompany the boat for an hour or so, keeping her course, occasionally swimming ahead or diving underneath from one side to another. Surely they are not trying to protect the vessel's hulk, which positively dwarves them. It seems far more likely that they do it out of playfulness.

Their developed social needs are likely responsible for a more depressing phenomenon, however. Of the dolphins that survive their capturing process, 53% die within the first three months of captivity and at least 50% of the ones that survive past that mark die in less than seven years. In the wild they can live for up to 40–60 years.[422] Supporters of dolphin therapy claim that only animals born in captivity are used for therapeutic purposes, but only two-thirds of marine mammals displayed in U.S. marine parks were born in captivity.[423]

Dolphin therapy is a rather loose term that covers a range of approaches. The simplest of these involves swimming and playing with dolphins. This approach is based on a premise that mere contact with the animal is sufficient to initiate a beneficial change in the patient. This kind of "therapy" rarely involves an actual therapist. The animal and the patient are mostly left to their own devices.

In a slightly more complex therapeutic format known as dolphin-assisted therapy (DAT) it is the therapist, or team of therapists, and not the animal, who is responsible for the treatment. The dolphin is there just to assist the process and to provide extra motivation for the patient to make an effort and go through prescribed exercises. This type of therapy uses structured programs tailored to the patient following his or her needs analysis. The child is encouraged to perform certain tasks, such as hanging a ring on a peg or uttering a particular word. When successful, the child is rewarded with a swim with the dolphin.

Another form of therapy is based on a belief that dolphins cure patients with therapeutic ultrasounds emitted by their echolocation center. This approach assumes that the dolphin "diagnoses" the patient, finds the affected tissue, and targets it with the right kind of ultrasound wave until it is cured. Proponents of this approach, known as the sonophoretic model in dolphin assisted therapy, also claim that ultrasound stimulation increases the flow of certain specific hormones that cross cell membranes improving cell-to-cell exchange of substances and the circulation of fluids. Moreover, they believe that dolphin echolocation has analgesic effects and that the contact with dolphins causes considerable changes in the bioelectrical activity of the human brain.[424]

Finally, a less popular hypothesis about the therapeutic effect of dolphins holds that the sounds emitted by these animals influence humans in ways similar to music therapy. In practice, these various approaches are often mixed with each other, producing the most common, eclectic approach where therapeutic objectives become inseparably linked with marketing goals.

These enthusiastic pictures of dolphin therapy, however, seem to be confined to the movement itself. Independent research reveals a different story altogether. In a rating published by Research Autism, a UK-based charity dedicated to the promotion of quality research into autism therapies, dolphin therapy scored the worst risk mark of three out of three and an additional question mark denoting mixed or limited evidence of the therapy's efficacy. This placed it in a group of the most dangerous therapies offered to autistic children. What did it to deserve this? Research Autism cited aggressive dolphin behavior leading to injury as one of the most dangerous threats to participants. The organization based this claim on available statistics of injuries suffered by participants of dolphin therapy.[425] Yes, the dolphin, that much-loved creature that is universally regarded as the epitome of gentleness, is in fact a highly aggressive predator.

> Three out of four attacks by bottlenose dolphins noted in recent weeks by volunteers from New Quay-based Cardigan Bay Marine Wildlife Centre were fatal. … The center suggested the attacks may be over competition for food or the result of dolphin mating behavior. …
> In May, volunteers at the center (CBMWC) in Ceredigion rescued a porpoise that stranded on the beach after being chased by dolphins. Last month researchers saw three dolphins killing a porpoise, with another similar incident a week later. Then last week three dolphins spent 20 minutes attacking a porpoise close to the centre's research vessel Anna Lloyd. They repeatedly pounced on the porpoise, forcing it underwater then throwing it in the air close to the boat.
> Researcher Milly Metcalfe said: "One of the dolphins in particular was attacking the porpoise while the others joined in from time to time. Although we were close by, they took no notice of us, intent on the attack." The crew brought the porpoise's body on board and found blood was coming from its mouth, suggesting it may have had punctured lungs or other internal injuries, she said.
> Ms Perry said the center had been left baffled by the recent spate of these attacks. She said: "One possibility is that they see the porpoises as competitors for food, especially if there's a shortage of prey in the area.
> "Although porpoises normally go for smaller prey, dolphins will eat anything. However there's been no other indication recently of any shortage."
> Another theory is low numbers of females is prompting the attacks by males. Male dolphins have been known to kill young dolphins in order to mate with the calf's mother. Porpoises are similar in size to a dolphin calf. The center said it is unclear which dolphins have been responsible for the attacks. Mr Perry added: "One animal we're familiar with, Nick, a female, was seen in the area when we rescued the porpoise last month, and she was also seen nearby in last week's attack. "Our records show that she's been seen before with the animals we think were responsible for the attack, so it may even be that she's teaching the others."[426]

Young males form alliances whose behavior resemble that of inner city gangs. One of their targets are female dolphins, which are abducted to release the males' sexual tension. The males are in a difficult position. Females only bear one young at a time and then spend between three and five years caring for their offspring. This means that sexually attractive females tend to be scarce. What is more, a single dolphin is powerless in the face of a strong female guarded by some experienced males. This lack of opportunity to release sexual tension often leads to acts of aggression and other surprising behavior in certain dolphins. Marine biologists researching populations of bottlenose dolphins observed that young males coped with the lack of sexually active females by rubbing their noses against their genitals and penetrating each other anally.[427] These homosexual behaviors are quite frequent and sometimes lead to humans falling prey to sexual assaults. Sexual frustration and aggression in dolphins, both males and females, has also been reported in dolphin therapy centers. Despite being carefully hidden by the owners, several such events have been documented.[428]

> In November 1999, the Bermudan Ministry of Environment reported that at least two people had been bitten during scheduled encounters swimming with dolphins in a facility in Bermuda. The bites were serious enough to require emergency hospital attention (Ministry of the Environment, Bermuda 1999). In August 2000, an 11-year-old child was bitten by a beluga she was interacting with at a marine park in Canada. The injury to her hand required stitches (Ananova 2000). In June 2003, a Japanese newspaper reported that a woman had sued a hotel in Taiji, Japan for injuries she sustained, including several broken bones in her ribcage and back, after a dolphin smashed into her during a swimming-with-dolphins encounter at the hotel (Mainichi Shimbun 2003). In 2006, a woman swimming with dolphins in a sea pen facility in Cuba was struck in the ribs by a captive dolphin and suffered a broken rib and a punctured lung (Stallard 2006).[429]

Samuels and Spradlin observed dominating behavior in bottlenose dolphins, the species most frequently used in therapy. They reported several types of aggressive behavior towards humans who stayed in the water with them, including: threatening with an open beak, nudging, forcing to back off, pressing against the patient, sudden turns, very fast maneuvers in direct proximity of humans, snapping, pushing, hitting with flippers, splashing the tail against the water, and sexual behavior.[430]

Many biologists who actively study and train dolphins point to the fact that dolphins can be aggressive even towards their trainers, i.e. persons with whom they have frequent and regular contact.[431]

A direct dolphin attack is not the only threat, however. Humans can contract dolphin diseases, mainly linked with parasites and bacteria, to which the human body is not immunized. In a certain Hawaii-based dolphinarium, scientists isolated fifteen types of bacteria, at least half of which were dangerous to humans. There were also numerous cases of fungi infections in veterinary doctors attending sick dolphins requiring difficult long-term therapies.

The risk of contracting animal diseases is much greater in captivity than in the wild. Bacteria that are harmless to dolphins but cause diseases in humans enter their bodies mostly through the respiratory tract or through cuts and wounds.[432] A study conducted in the USA on people who had systematic contact with marine mammals showed that 50% of them were wounded by these animals and 23% would develop a rash or other allergic reaction after contact.[433] This was possible because, despite the fact that captive animals are kept in chemically controlled environments, they continue to breathe, urinate, and defecate in the direct vicinity of recipients of animal therapies. This heightens the risk of brucellosis, which has a particularly devastating effect on the human body, while in pets it causes miscarriage and infertility.[434] Dead marine mammals found washed on the beach are often diagnosed with an infection of *Brucella*.[435] The same bacteria have been isolated in captive dolphins.[436]

The American Veterinary Medical Association acknowledged this problem in its 2007 guidelines: "A wellness program should be instituted for animals participating in AAA, AAT, and RA programs to prevent or minimize human exposure to common zoonotic diseases such as rabies, psittacosis, salmonellosis, toxoplasmosis, campylobacteriosis, and giardiasis. Need for specific screening tests should be cooperatively determined by the program's attending veterinarian(s) and physician(s)."[437]

This recommendation has not been followed and dolphin therapies around the world are not required to comply with any such regulations. This means that centers offering these services are not required to meet any specific hygienic standards which could protect them from contracting such diseases.

While we may be especially concerned with human health, animals are in no lesser danger of contracting diseases. Dolphins can also develop sicknesses to which they are not resistant. In captivity they die quickly and their population dwindles.

The cost of such therapy is another factor worth mentioning, especially when compared to the effect. Until recently, one of the most expensive was the now defunct Dolphin Human Therapy, which in 2006 cost 7,850 dollars for two weeks, or 11,800 dollars for three weeks. This price was exclusive of travel and accommodation of the child and its guardian.

The German association Autismus followed Research Autism in speaking out about dolphin therapy. In 2006, it officially stated that there was no convincing evidence for the efficacy of dolphin therapy in treating autism.[438]

One year later, the Whale and Dolphin Conservation Society published a large report summing up various doubts about the efficacy of dolphin therapy, the ethical aspects of using the animals for this purpose, and the risks to both the participants and the animals.[439] The report revealed uncomfortable facts and the kinds of details that very rarely make it through to the mainstream media, as well as some that were outright hidden by dolphinaria and dolphin therapy centers. The authors confirm the risks to people in contact with the animals covered earlier in this chapter. A large proportion of the document is devoted to an analysis of risks to animals from contact with humans. It also provides a careful analysis of various therapies and, perhaps most importantly, offers a review of available research into the efficacy of dolphin therapy and of comparative studies of dolphin therapy vs. other therapies. It is these types of studies, including ones not covered by the report, that we now turn our attention to.

The literature on the subject abounds in anecdotal evidence of the effects of dolphin therapy, but disproportionately few empirical studies are available about its efficiency. What is available, however, are a number of robust comparisons of such research. In one of these, published in 2003, Tracy L. Humphries analyzed six key studies on the efficacy of dolphin therapy on disabled children aged six or younger. She concluded: "Despite intensive research, there is still no scientific proof that DAT helps curing autistic children. To help them, the association strongly recommends a lovingly consistent education, behavioural therapy including preparation of speaking and social training. Dolphins are wild animals. It's a matter of ethics, whether we are allowed to deprive them of their freedom to exploit them for questionable purposes."[440]

Lori Marino and Scott Lilienfeld published another broad review. They looked at research published between 1999 and 2005 and found five empirical studies. All of these were found to suffer from serious methodological deficiencies and a large number of smaller problems in terms of both internal and external validity. What Marino and Lilienfeld singled out as the most serious problem was a lack of control over other factors involved that could have an impact on the therapy outcomes. The very fact of being in water, being engaged in movement activities linked with strong stimulation, and the contact with the therapist and the experimenter effect could all have an impact on any change observed after the therapy. A well-designed study should include a control group consisting of children involved in similar exercises in water, but without dolphins. Lori Marino and Scott Lilienfeld ended their review by saying that every single conclusion made in these studies was dubious at best and entirely unfounded at worst.[441]

A team of German researchers managed to avoid most of the methodological pitfalls mentioned above in a piece of research they published a few years after the Marino and Lilienfeld review. They worked with four groups of children who shared similar socialization and speech problems. The experimental group took part in dolphin therapy activities in a relaxing and therapeutic environment. Consulting sessions for the parents were offered to augment the activities. The first control group of children only got to play with dolphins, while the second control group engaged in the same activities as the experimental group, but farm animals replaced the marine mammals. This second control group, however, had not been adequately selected, which crippled its comparability, a problem acknowledged by the authors themselves. The third control group did not take part in any activities. The strongest therapeutic effect was observed in the experimental group. The isolated contact with dolphins in the first control group produced only minor improvements in communication competences and no therapeutic effect was identified in the therapy involving farm animals. The authors concluded that it was the participation of the whole family, their understanding of the relationships, etc., that provided the greatest contribution to the therapeutic outcome. Conversely, the fact that the benefit from just playing with dolphins was minor and that it was independent from whether the therapy took part in a marine environment or in a closed dolphinarium led the authors to conclude that the mystical claims about the ultrasonic effect had no empirical grounds.[442]

In her paper published in 2008, Cathy Williamson of the Whale and Dolphin Conservation Society, UK, expressed a highly critical opinion of dolphin therapy similar to that of Research Autism covered earlier. Williamson maintained that dolphin therapy was not just ineffective, but also dangerous to the health of both the people and animals.[443]

Among the numerous claims made by proponents of dolphin therapy the most fantastical is their hypothesis about the therapeutic effect of ultrasounds emitted by these animals. This claim combines the credible with the supernatural. The hypothesis about the therapeutic effect of dolphin-emitted ultrasounds in founded on a number of medical applications of ultrasound waves, whether already established, such as breaking down kidney stones and destroying fatty cells as part of liposuction, or still under development, including the high intensity focused ultrasound (HIFU) method currently tested for arresting internal hemorrhage or eliminating certain types of cancer. However, supporters of this approach in dolphin therapy go further and assume that dolphins have a supernatural power of discovering sick tissue, an ability to precisely concentrate their ultrasound waves on it, and accurately select the correct strength needed to cure it. Where does this idea come from? Perhaps these undoubtedly intelligent mammals that constantly move and frolic in the water have mastered what scores of specialized researchers find so difficult despite being equipped with state-of-the-art laboratories and precision metering devices?

These are exactly the type of questions tackled by a team of researchers from the Institute of Behavioral Research at Free University, Berlin, Germany. Karsten Brensing, Katrin Linke, and Dietmar Todt analyzed dolphin behavior observed during 83 therapeutic sessions in the popular center Dolphin Plus in Florida Keys, USA. Since the previously described medical and physiotherapy studies concluded that the efficacy of ultrasound treatment required specific levels of power, exposure, time and number of repetitions, the team investigated whether dolphin behavior offered these levels of ultrasound activity understood as a precondition for the success of this kind of "treatment." The study found that not a single dolphin included in the survey met the requirements at even a minimum level. Therefore the researchers concluded that there were no grounds for confirming hypotheses made at an initial stage in terms of human treatment with dolphin-emitted ultrasounds.[444]

A description of the efficacy of dolphin therapy would not be complete without mentioning the surprising results obtained by David Nathanson, one of the leading champions and authorities on dolphin therapy. In his research he employed an artificial dolphin known as *Therapeutic Animatronic Dolphin* (TAD) supplied by Animal Makers, specialists in developing electronically controlled animal dummies for the film industry. In his work Nathanson included 35 children from seven countries with various dysfunctions and ailments ranging from autism to cerebral palsy, Rett syndrome, Lennox-Gastaut syndrome, dystrophy, etc. The results were nothing less than astounding, and this is what the author had to say about them:

> First, no significant difference between dolphins and TAD was found in eliciting orienting responses of touching and/or saying words for all study participants. Second, no significant difference between dolphins and TAD was found in eliciting orienting response of touching and/or saying words for children within ability groupings of moderate or severe levels of disability. For children with profound disabilities, TAD was significantly more effective in eliciting an orienting response of looking. Third, response times to either dolphins or TAD were the same whether the reinforcement was given from a platform or in water. Interaction with TAD provided the same or more therapeutic benefits as interaction with dolphins, without the environmental, administrative/legal and practical limitations, including high cost, associated with dolphins.[445]

Both this study and his earlier work suffered from several methodological problems, such as a non-homogenous experimental group, lack of a control group, etc. For these reasons his experiment needs to be replicated in more stringently controlled conditions for confirmation. Nevertheless, the study must be seen as pioneering in this area because it offers a way to isolate factors of primary significance in dolphin therapy. If this would be made to work then perhaps the precisely controlled contact with water, the novelty effect, positive emotions, movement, and other factors might at last deliver predictable effects where previously they were occasional and unpredictable, as is the case with contact with real dolphins. Additionally, the approach could be freed from all risks associated with dolphin therapy, including risks to animals.

While the efficacy of therapies involving dolphins has not been scientifically demonstrated there are positive accounts of parents of children who benefited from dolphin therapy. This kind of information must never be ignored, and so researchers have been trying to identify the real causes of positive change observed in the children's self-reported wellbeing, new skills, new behaviors, or greater motivation for further work. Lori Marino and Scott Lilienfeld addressed these questions and concluded that the reason may be found in certain well-known effects that are not necessarily related to the presence of dolphins during the therapy.

The well-documented placebo effect is the first potential culprit. It occurs when the person subject to therapy and/or people around that person are expecting an improvement and are assuming that it is under way. Indeed, considering how widely advertised dolphin therapy is and the lengths intermediaries and delivering organizations go to in fueling the hopes of potential patients and their guardians, it would be hard to imagine such expectations not arising. Perhaps this in itself might be sufficient for some actual beneficial effects to occur. This could work in the same way as the visit of a doctor or a sufficiently long conversation with one can sometimes lead to a major improvement in the patient's condition.

The second effect named by Marino and Lilienfeld is the novelty effect. For most people new events and experiences tend to be exciting, bring new hope, or improve motivation. Attractive surroundings and a holiday atmosphere where daily life seems far away may augment this effect. This is exactly what almost always happens during dolphin therapies, while there still is that additional factor of the presence of a large and peculiar animal. A new situation that is exciting and instills enthusiastic responses may be a desirable stimulus for change.

The authors dubbed the third of their effects construct confounding. This effect occurs when researchers or observers of a phenomenon fail to take into account that the therapeutic procedure they are observing may engage an entire set of active factors rather than just the one they are interested in. In dolphin therapy interaction with the animal is accompanied by a complex set of stimuli, including: being in the water, being the focus of attention of therapists, often being in an exotic environment, and experiencing a very different form of therapy to that enjoyed before. On the one hand, it is exceedingly difficult to design a study that would isolate individual factors and identify which ones, acting alone or in concert, have a significant influence on therapy outcomes. On the other hand, it is very easy to attribute effects observed to just the single variable that is followed, i.e. interaction with the dolphins.

Alongside these three effects, which are found in virtually every observed dolphin therapy, the authors found numerous others that were less frequent. These included: demand characteristics, i.e. a tendency of participants to alter their responses in accord with their suspicions about the research hypothesis; experimenter expectancy effects, which is a tendency for the experimenter to unintentionally bias the results in accordance with the hypothesis; informant bias, the tendency of informants to selectively recall improvement in accord with their hopes and expectations (retrospective bias) or unintentional distortion of improvement due to effort justification; and multiple intervention interference, i.e. administration of treatments other than the intended treatment during the course of the study and so forth.

I am fully aware that to the parents of disabled children all these research results, even the most striking ones that have been gathered in carefully designed scientific experiments, observations, and analyses, amount to nothing in the face of accounts of other parents who are convinced that their children experienced a "miracle." It is entirely understandable that any hope is more alluring to them than brutal facts offered by scientists. Therefore, the facts reviewed in this chapter may have failed to dissuade the reader from a decision to devote their own financial resources, time, and energy, as well as that of their child's, to participate in dolphin therapy. If you fall into this category, I have one last suggestion. Before you pack your bags for the trip, ask the parents who have recommended dolphin therapy to meet with you once again and ask them these questions:

- Did your child undergo any other animal-assisted therapy before dolphin therapy?
- If so, what were the effects?
- Did these effects differ significantly from the dolphin therapy effects?
- Did your child undergo any kind of pool-based therapy supervised by a qualified therapist before dolphin therapy?
- If so, what were the effects?
- Did these effects differ significantly from the dolphin therapy effects?
- What specific changes were caused by dolphins?
- Which change and progress would not have been possible in any other way than dolphin therapy, in your opinion?
- How persistent were these effects?
- What was different in what dolphins and dolphin therapists did to what previous therapists did?
- Was there anything worrying that occurred? What was it?

---

[420] http://dolphin-therapy.org/general-therapy-dolphins/

[421] http://www.waterplanetusa.com/dolphin-therapy/

[422] http://www.dolphins-world.com/wild-dolphins/

[423] See for more: http://www.sun-sentinel.com/sfl-dolphins-parksmay16-story.html#page=1

[424] K. Brengsing, K. Linke, and D. Todt, "Can Dolphins Heal by Ultrasound?" *Journal of Theoretical Biology, 225*, (2003): 99–105.

[425] http://www.researchautism.net/interventions/64/dolphin-therapy-and-autism/Risk%20and%20Safety

[426] http://www.bbc.com/news/uk-wales-28289877

[427] R. C. Connor, R. S. Wells, J. Mann, and A. J. Read, "The Bottlenose Dolphin: Social Relationships in a Fission-Fusion society," in: *Cetacean Societies: Field Studies of Dolphins and Whales*, eds. J. Mann, R. C. Connor, P. L. Tyack, and H. Whitehead, (University of Chicago Press, 2000).

[428] E.g.: https://www.youtube.com/watch?feature=player_embedded&v=jYLkSvf2VDc

[429] P. Brakes and C. Williamson, *Dolphin Assisted Therapy: Can you put your faith in DAT?* (Whale and Dolphin Conservation Society: October 2007), 4.

[430] A. Samuels, and T. Gifford, "A Quantitative Assessment of Dominance Relations Among Bottlenose Dolphins," *Marine Mammal Science, 13*, (1997): 70-99.

[431] T. G. Frohoff, "Stress in Dolphins," in: *Encyclopedia of Animal Behavior*, ed. M. Bekoff, (Westport, Connecticut: Greenwood Press, 2004); R. H. Defran, and K. Pryor, "The Behavior and Training of Cetaceans in Captivity," in: *Cetacean Behavior:*

Mechanisms and Functions, ed. L. M. Herman, (New York: John Wiley and Sons, 1980); J. C. Sweeney, "Marine Mammal Behavioral Diagnostics," in: *CRC Handbook of Marine Mammal Medicine: Health, Disease and Rehabilitation*, ed. L. A. Dierauf, (Boston: CRC Press, 1990).

[432] C. D. Buck, and J. P. Schroeder, "Public Health Significance of Marine Mammal Disease," in: *CRC Handbook of Marine Mammal Medicine: Health, Disease and Rehabilitation*, ed. L. A. Dierauf, (Boston: CRC Press, 1990); I. A. P. Patterson, "Bacterial Infections in Marine Mammals," in: *Zoonotic Diseases of UK Wildlife*. (Bath: BVA Congress, 1999).

[433] J. A. Mazet, T. D. Hunt, and M. H. Ziccardi, "Assessment of the Risk of Zoonotic Disease Transmission to Marine Mammal Workers and the Public: Survey of Occupational Risks. Final Report Prepared for United States Marine Mammal Commission," Research Agreement Number K005486-01, (2004).

[434] M. Tachibana, K. Watanabe, S. Kim, Y. Omata, K. Murata, T. Hammond, and M. Watarai, "Short Communications: Antibodies to *Brucella* spp. in Pacific Bottlenose Dolphins from the Solomon Islands," *Journal of Wildlife Diseases*, 42(2), (2006): 412-414.

[435] G. Foster, K. L. Jahans, R. J. Reid, and H. M. Ross, "Isolation of *Brucella* Species from Cetaceans, Seals and an Otter," *The Veterinary Record*, 138, (1996): 583-586; J. Maratea, D. R. Ewalt, S. Frasca, J. L. Dunn, S. De Guise, L. Szkudlarek, D. J. St. Aubin, and R. A. French, "Evidence of *Brucella* sp. Infection in Marine Mammals Stranded Along the Coast of Southern New England," *Journal of Zoo and Wildlife Medicine, 34* (2003): 256–261.

[436] Tachibana et al., "Short Communications."

[437] American Veterinary Medical Association, "Guidelines for Animal Assisted Activity, Animal-Assisted Therapy and Resident Animal Programs," (March, 2003), https://www.avma.org/KB/Policies/Pages/Guidelines-for-Animal-Assisted-Activity-Animal-Assisted-Therapy-and-Resident-Animal-Programs.aspx#references

[438] http://www.marineconnection.org/campaigns/dat_alternatives2006.htm

[439] P. Brakes and C. Williamson, *Dolphin Assisted Therapy: Can you put your faith in DAT?* (Whale and Dolphin Conservation Society: October 2007).

[440] T. L. Humphries, "Effectiveness of Dolphin-Assisted Therapy as a Behavioral Intervention for Young Children with Disabilities," *Bridges: Practice-Based Research Synthesis, 1*, (2003): 1–9.

[441] L. Marino and S. O. Lilienfeld, "Dolphin Assisted Therapy: More Flawed Data and More Flawed Conclusions," *Anthrozoos, 20*, (2007): 239-249.

[442] E. Breitenbach, E. Stumpf, L.V. Fersen and H. Ebert, "Dolphin-Assisted Therapy: Changes in Interaction and Communication Between Children with Severe Disabilities and Their Caregivers," *Anthrozoös, 22*, (2009): 277-289.

[443] C. Williamson, "Dolphin Assisted Therapy: Can Swimming with Dolphins be a Suitable Treatment?" *Developmental Medicine & Child Neurology, 50*, (2008): 477.

[444] Brensing et al., "Can Dolphins Heal."

[445] D. E. Nathanson, "Reinforcement Effectiveness of Animatronic and Real Dolphins," *Anthrozoos, 20*, (2007): 181-194, 181.

Chapter 15: Ad Infinitum

From the murky waters of the swamps near a place called Lerna, the hydra would rise up and terrorize the countryside. A monstrous serpent with nine heads, the hydra attacked with poisonous venom. Nor was this beast easy prey, for one of the nine heads was immortal and therefore indestructible.
Hercules set off to hunt the nine-headed menace, but he did not go alone. His trusty nephew, Iolaus, was by his side. Iolaus, who shared many adventures with Hercules, accompanied him on many of the twelve labors. Legend has it that Iolaus won a victory in chariot racing at the Olympics and he is often depicted as Hercules' charioteer. So, the pair drove to Lerna and by the springs of Amymone, they discovered the lair of the loathsome hydra.
First, Hercules lured the coily creature from the safety of its den by shooting flaming arrows at it. Once the hydra emerged, Hercules seized it. The monster was not so easily overcome, though, for it wound one of its coils around Hercules' foot and made it impossible for the hero to escape. With his club, Hercules attacked the many heads of the hydra, but as soon as he smashed one head, two more would burst forth in its place! To make matters worse, the hydra had a friend of its own: a huge crab began biting the trapped foot of Hercules. Quickly disposing of this nuisance, most likely with a swift bash of his club, Hercules called on Iolaus to help him out of this tricky situation.
Each time Hercules bashed one of the hydra's heads, Iolaus held a torch to the headless tendons of the neck. The flames prevented the growth of replacement heads, and finally, Hercules had the better of the beast. Once he had removed and destroyed the eight mortal heads, Hercules chopped off the ninth, immortal head. This he buried at the side of the road leading from Lerna to Elaeus, and for good measure, he covered it with a heavy rock. As for the rest of the hapless hydra, Hercules slit open the corpse and dipped his arrows in the venomous blood.[446]

Many years ago, when I began work on a trilogy dedicated to abuses in and of psychology, my objective was not to formulate an exhaustive lexicon of psychological pseudosciences and pseudotherapies. Such a task would be too much for just one man, and the Hydra of psuedoscience shares with the Lernaen Hydra one crucial characteristic: it is constantly growing new heads in every direction. I have also yet to succeed in locating the strength of Hercules within myself. If I desired to create such a lexicon, I would have to write at least ten volumes, in the knowledge that it would still not be enough, for new pseudotherapies are constantly emerging. A never-ending story...
Writing the section on child therapies, I wanted to demonstrate using a few examples of popular pseudosciences the mechanisms by which they arise and operate, while also desiring to analyze the reasons for their popularity. The Doman-Delcato method and educational kinesiology are excellent for this purpose considering their prevalence. Attachment therapy, while less popular, demonstrates the limits to which child therapists can push the

boundaries, and also shows the mechanisms that aid them in maintaining their impunity. During the course of work on these pseudotherapies, I have come across so much information about ineffective, dangerous, and highly suspect concepts that I felt I should take the opportunity to write about them here, by way of warning.

While writing about the Doman-Delacato concept, I consulted with doctors and child rehabilitation experts, who expressed very strong criticism of therapy conducted on the basis of that method; at the same time, they encouraged me to take an interest in other, equally suspect, methods of treating children with brain damage. The name which most frequently cropped up in that context was the *Vojta Method*, named after the Czech doctor who developed it, Václav Vojta. According to Internationale Vojta Gesselshaft e.V., this is a therapy applied in the rehabilitation of children and adults with neurological disorders in Germany, Chile, Japan, Korea, Mexico, Norway, Austria, Poland, Romania, Spain, the Czech Republic, Italy, Colombia, Peru, Syria, and Taiwan.[447] More research reveals that it is also applied in the United Kingdom,[448] and as we can read on the web page of New York Dynamic Neurmuscular Rehabilitation & Physical Therapy: "In Europe Vojta therapy has been the method of choice for the treatment of Cerebral Palsy, motor delays, strokes, multiple sclerosis and other neurological disorders and only now it is making its way to the United States."[449]

Vojta therapy is most frequently described as a "neurophysically directed system of priming," and is designed to restore inborn patterns of locomotor movement whose development has been retarded by early-age brain damage, or which has been lost as a result of injury. This method – also referred to as reflex activation– is applied both as a preventative and a therapeutic treatment in pediatric motorneurological disorders and orthopedic posture deficits. It is based on the principle of evoking reflexes: pressure placed on specific points on the body (stimulus points) evoke the kinds of reflexes that emerge in the course of normal motor development at its later stage (e.g. rotation). In this manner, pathological posture and movement patterns can be corrected. Vojta therapy applies in particular two "priming" systems to adjust posture: reflex creeping and reflex rolling. The method is designed for infants and toddlers, and can therefore be used during the first and second years of children's lives, when the improper reflexes have not yet taken root. The key exercises of reflex creeping and reflex rolling, whose movement sequences are precisely defined in advance, are supposed to prevent the development of improper movement sequences and to reinforce correct ones. These reflexes are evoked by the therapist (parents) applying pressure to several points on the body, primarily the head. The child thus stimulated in several key points is moved into an unpleasant position, and attempts to escape it. The only means by which the child may free itself from this uncomfortable position it is being held in by force is to perform one particular movement, the one expected by the therapist and which constitutes the desired sequence. One session of such exercises lasts around twenty minutes and should be repeated three to four times per day.

This rehabilitation technique has been subjected to fierce criticism. Opponents of the method argue that such young children are incapable of understanding the purpose of performing such exercises, but they understand perfectly well that something bad is being done to them, and that by their own parents. This is illustrated by the opinions of mothers that can be found on the Internet:

> Vojta – awful, just awful – I couldn't get my 6-year-old daughter to do the exercises, she just broke down in tears when I told her it was time to start (and to be honest, I wanted to as well). It was also really hard to find time to do the exercises several times a day. We started rehabilitation when she was 2 months old, using the NDT-Bobath method (3 times a day for 20–30 min.), but our son quickly started crying, and he even began to sob when we placed him on his back, he wouldn't calm down until we cradled him in our arms. It's even worse now with the Vojta method, the little guy starts crying when we begin undressing him, and he can't calm down even long after the exercises are done. He's generally upset, he can get scared just by someone making sudden movements or noise. Today he cried in his sleep for the first time. So my question is, if the Vojta method or NDT-Bobath makes my child fearful and leads him to tears, won't it have a negative effect on his psyche?[450]

Naturally it will! Setting aside the issue of the method's effectiveness in achieving the objectives it was created for, there can be no question of its capacity to evoke serious anxiety. Indeed, conditioned anxiety is the real result guaranteed by the application of the Vojta method! The question remains of which stimuli that reaction will be mapped to. In the best case, although highly doubtful, it will remain associated with external circumstances: the building or room where the rehabilitation is conducted. What is more likely, however, is that the child will begin to react with anxiety upon seeing the individuals responsible for leading the therapy, such as the parents in the case described above. This implies that an integral method of the Vojta therapy should be a procedure for eliminating anxious reactions.[451] Yet I have never encountered such recommendations.

In the course of reviewing extensive databases containing medical and psychological scholarly articles (EBSCO and PubMed), I came across just around two dozen articles dedicated to the Vojta method. One of them engaged the issue of the negative psychological impact the therapy may generate. Its authors attempted to answer the question of whether the Vojta therapy led to greater stress and more negative results than Bobath therapy. Indeed, doubts are already raised by the very concept of the research; as we have seen in the words of one mother quoted above, the Bobath method also evokes negative effects. In this sense, it is not a reliable control group which could be used in comparisons to generate trustworthy results. It is tantamount to comparing a group of physically abused children and a group of sexually abused children when answering the question of what leads to more dramatic psychological effects. In spite of these errors, the work does cast light on the issue of mother-child relations in therapy conducted via the Vojta method. It is worth recalling the researchers' determination that the Vojta method causes far greater stress and psychological burden at the inception than the Bobath method, but in the course of the therapy these differences disappear. The authors of the article appeal for families beginning to apply the therapy to be informed of the potential psychological consequences. They perceive the necessity of providing psychological support in the course of the therapy[452]. However, physiotherapists employing the Vojta method do not take this issue as seriously: "The therapeutically desired activated state often expresses itself during treatment in new-born babies as crying. This understandably leads to parents feeling concerned, and makes them assume that it is 'hurting' their child. At this age, however, crying is an important and appropriate means of expression for the little patients, who react in this way to unaccustomed activation. As a rule, after a short familiarization period, the crying is no longer so intense, and in breaks from exercise as well as after the therapy, new-born babies calm down immediately."[453]

To recommend a method whose consequences include the systematic and daily experience of suffering, one must possess an overwhelming certainty that the positive effects of the therapy are sufficient to counterbalance its negative ones. Do those employing the Vojta method display such a conviction? Unfortunately, they do! The same conviction as the majority of the cargo cult priests. A conviction built without a foundation, resting on the pillars of mere opinion, enhanced with thin anecdotal evidence; there is precious little empirical proof to be found. The majority of the scientific works that I have found and written about appeared in journals at the margin of the field. Some of them were quite odd, such as one in which the authors compared the areas stimulated by Vojta therapy with those by acupuncture, followed by their call to merge the two systems.[454]

Taking a closer look at what supporters of the therapy write also fails to provide us with any certainty as to the strength of its foundations. For example, in describing the method they write: "This took place during the 1950s and 1960s. It was a period of intense scientific exploration conducted around the world, in a search for an effective means of helping people with damage to the central nervous system resulting in motor disorders. During that period, a number of neurophysiological rehabilitative concepts arose that remain well-known and in use until the present day, such as NDT-Bobath, PNF, and Doman-Delacato."[455]

Just the mere mention of the Vojta method in the same breath as other "well-known and successfully used" methods such as Doman-Delacato sets off alarm bells. As it is, the methods share some significant traits. Physiotherapists implementing the Vojta method, similarly to supporters of Doman and many other child therapists, place strong emphasis on the responsibility of parents for results: "The key to successful rehabilitation using the Vojta method are *PARENTS*, who are responsible for it being pursued systematically. The physiotherapist's task is to solve technical issues arising in the course of treatments. *A necessary condition of achieving maximum effect is the awareness of parents that their child has a developmental problem and their conviction that initiating and following through with therapy is vital.* The Vojta therapy is not among the easier ones."[456]

It would appear that the field of child therapy is deeply infected with pseudoscience. I suspect that a more critical review of the therapies mentioned in the same breath (NDT-Bobath, PNF, Doman-Delacato, etc.) could generate quite interesting results.

Meanwhile, Vojta therapy has been transformed by the Czech pediatrician Pavel Kolar into the Dynamic Neuromuscular Stabilization method, which is emerging as one of the standard physiotherapy treatments.[457] Whether there is any pseudoscience, and if what is the extent of it, are questions which can only be answered following very thorough study and analysis.

The Research Autism referred to in the preceding chapter lists among unsafe therapies primarily those employing pharmacological substances and/or supplements. However, their list does include one more form of psychotherapy which has received the maximum three points on the danger scale, and also the maximum possible score in the "very strong evidence disproving the method's effectiveness" scale, which positions it as the undisputed leader of dangerous therapies. This technique is called *facilitated communication* (FC), and it was developed in the late 1980s by the Australian doctor Rosemary Crossley, who sought a means of supporting people with autism in communicating with their surroundings. It consists in holding the hand, forearm, or arm of the patient by a facilitator and writing with the assistance of a keyboard or a special "simplified communication board." Autism is frequently accompanied by difficulties with praxis, the planning of complicated target movements. To overcome these difficulties, the creator of the method claims that physical assistance in the form of human contact should be applied.

Why has FC come in for such sharp criticism? After a period of tremendous enthusiasm at the beginning, studies were undertaken of its effectiveness and credibility. This occurred contemporaneously with a wave of court cases concerning claims of sexual abuse.[458] During these trials, statements were often used that had been acquired with the assistance of FC. Their authenticity was frequently brought into question, and authorship of texts generated in this manner is today assigned to the therapist rather than the patient. Additionally, texts arising during the course of therapy sessions frequently are so illegible as to require interpretation, which is generally performed by a therapist possessing no scientifically verified guidelines for such interpretation.[459]

In 1992, the American Academy of Pediatric Committee on Children with Disabilities published a statement, which reads in part: "In the case of FC, there are good scientific data showing it to be ineffective. Moreover, as noted before, the potential for harm does exist, particularly if unsubstantiated allegations of abuse occur using FC. Many families incur substantial expense pursuing these treatments, and spend time and resources that could be used more productively on behavioral and educational interventions. ... Until further information is available, the use of these treatments does not appear warranted at this time, except within research protocols."[460]

The method has also come under fire from the American Association on Mental Retardation, the American Psychiatric Association, and the American Academy of Child and Adolescent Psychiatry. Systematic reviews of studies lead to the conclusion that there is no evidence verifying the effectiveness of FC. Quite the opposite – everything would seem to indicate that it is an ineffective procedure, without any justification for its use.[461]

A review of studies composed by a pair of Polish authors contains this assessment: "It can therefore be concluded that evidence for the effectiveness or lack thereof of FC is closely linked to the manner in which research is conducted. The results of works taking control procedures into account do not confirm the effectiveness of FC; studies with smaller control groups generate inconclusive results, and those ignoring standard control procedures generally demonstrate that FC is an effective method."[462]

In an article titled "Evidence-Based Practice in Psychology and Behavior Analysis," William T. O'Donohue and Kyle E. Ferguson say directly that the application of problematic therapeutic interventions – such as FC – is a crime in the case of autistic children.[463]

Despite such categorical assessments of this form of communication, in 2010 the world's media reported on the following sensation:

> A car crash victim has spoken of the horror he endured for 23 years after he was misdiagnosed as being in a coma when he was conscious the whole time.
> Rom Houben, trapped in his paralysed body after a car crash, described his real-life nightmare as he screamed to doctors that he could hear them – but could make no sound.
> "I screamed, but there was nothing to hear," said Mr Houben, now 46, who doctors thought was in a persistent vegetative state.
> "I dreamed myself away," he added, tapping his tale out with the aid of a computer.
> Doctors used a range of coma tests before reluctantly concluding that his consciousness was "extinct."
> But three years ago, new hi-tech scans showed his brain was still functioning almost completely normally.
> Mr Houben described the moment as "my second birth." Therapy has since allowed him to tap out messages on a computer screen.

Mr Houben said: "All that time I just literally dreamed of a better life. Frustration is too small a word to describe what I felt."[464]

In fact, Ron Houben "said" all of this to his therapist Linda Wouters, who applied FC, and the "statements" thus recorded were to be hard evidence of the patient's functional cognizance, at least in the opinion of his doctor, Steven Laureys. Sometime after, the circumstances surrounding this "miracle" were subjected to thorough examination and escribed by Belgian skeptics. A simple test was conducted during the study. Without the participation of Anne C., who had by that time assumed the role of his therapist, Houben was shown pieces of paper on which words were printed in large letters, and the words were read aloud several times. Next, the therapist was brought in, who applied the facilitated communication method to help her patient describe which word he had just heard or seen. Unfortunately, all of the answers were incorrect. This procedure was repeated multiple times in an array of configurations. Words were replaced with images. The results went unchanged. Not a single response given by Houben was correct. Every attempt resulted in failure. Ron's consciousness, it seemed, only began to function when the relevant knowledge was possessed by his therapist.[465]

Interestingly, and fitting of an archetypical pseudoscience, adherents of FC have found a manner to discredit studies generating results that conflict with their interests. They declare that subjecting autism patients to assessment during the application of FC leads to a loss of trust and confidence in their own possibilities, and in turn of their capacity to communicate with others. As a consequence, motivation to engage in monitored and assess communication is lessened, which explains why the studies do not generate the same results that therapists achieve during individual work with the patient.[466]

Perhaps if I were to hear something like that for the first time, I could treat it seriously. The thing is, for decades people convinced of the existence of various paranormal phenomena – including parapsychological capabilities – have repeated similar explanations like a mantra. The presence of individuals skeptical towards parapsychological phenomena (here we may freely replace "parapsychological" with "facilitated communication," "hypnotic regression," "reincarnation," etc.) inevitably interrupts their occurrence, in the same way as measuring them does. If this is the case, then we ought to give consideration to the hypothesis posited by professor Łukasz Turski during a discussion with a parapsychologist, who declared that he accepts such an explanation for the "disappearance" of the phenomena being studied, while at the same time stating that he is so extremely skeptical towards such phenomena as to disrupt the recording of measurements in all parapsychological experiments conducted around the world.[467]

FC has led to the disintegration of many families, has harmed innocent people, and it is more than doubtful whether its positive effects have balanced out even a portion of the damage it is responsible for.

During conversations with parents of children with developmental disabilities, I have frequently been asked to give my opinion on the purportedly wonderful *Tomatis Method*. This is one of the methods categorized under the broad label of auditory integration training (AIT). The fundamental assumption underlying the majority of versions of this training is the statement that proper auditory stimuli have healing properties. The experts from Research Autism have awarded AIT two (out of a possible three) points on the "strong evidence against the effectiveness of the method."

Alfred A. Tomatis maintained that sound messages are usually correctly heard by people but poorly analyzed in an emotional framework. The brain protects itself by constructing barriers that can result in the development of various disorders. To overcome these barriers, he experimented with a mother's voice as heard in the uterus. Tomatis played recordings to his patients of their mother's voices to treat a variety of disorders, including dyslexia, autism, and depression. The electronic ear, he maintained, could simulate the sound of the mother's voice and lead the child gradually to accept and respond to her real unfiltered voice. He reported that this method often brought startling results, with children crying with joy as they recognized their mother's voice for the first time.

What is striking in the marketing of such therapies as Tomatis's method is the argumentation focused on one factor as a dominant cause of many problems such as autism, schizophrenia, depression, dyslexia, and many, many more. Moreover, as the marketers persuade potential patients, addressing that one factor leads to a happy, healthy, fulfilling life. This marketing formula appears well established among proponents of pseudoscience and it goes far beyond the available evidence.

And what does the evidence say? The well-known clinical neurologist and dedicated skeptic Steven Novella reviewed the available research results. He came to the conclusion that proponents of the Tomatis method:

> Summarize the results of one study for auditory processing disorders, three for anxiety, and four for learning disabilities and behavioral problems (one was a meta-analysis of five studies with a total of 225 subjects). That's it, and they did not provide references.
>
> The studies are all small and uncontrolled or not blinded, many are just case series where Tomatis practitioners followed the progress of their clients. In other words – this is very low grade evidence, barely above testimonials, and completely unreliable, especially since many relied upon subjective outcomes.

I could not find much in a PubMed search to support auditory integration training or the Tomatis Method specifically. I did find one review of auditory integration for autism spectrum disorder (ASD) which concluded that "There is no evidence that auditory integration therapy or other sound therapies are effective as treatments for autism spectrum disorders. As synthesis of existing data has been limited by the disparate outcome measures used between studies, there is not sufficient evidence to prove that this treatment is not effective."[468]

At present there is insufficient evidence to justify using the therapy for ASD. "I did find preliminary evidence for altered auditory processing in those with panic disorder. It is not clear, however, if this contributes to the anxiety or is simply a result of the anxiety, or if both symptoms are a result of an underlying disorder. It is a huge leap to conclude that training auditory processing will affect the anxiety."[469]

Nothing more concerning auditory integration training generally, and the Tomatis method in particular, has been published and indexed in PubMed since Novella wrote his article. If parents of disabled children would like them to sacrifice part of their kids' childhood to engage them in an experiment started by Alfred Tomatis, they should be aware that it is just that: an experiment.

Even more popular than auditory integration training is *sensory integrative therapy* (SIT, also known as sensory integration training). Its roots go back to the 1960s, when it was developed by the therapist Jean Ayres from Los Angeles. It is applied to children with cerebral palsy, ASD, Down syndrome. SIT is based on the idea that some people have difficulties in receiving, processing, and making sense of information provided by their senses. For example, some people with autism are hyper-sensitive to some stimuli such as loud noises, but hyper-sensitive to other stimuli such as pain. In SIT, the therapists assess a person's sensory difficulties and then develop a personalized treatment program in which they use the most appropriate techniques and tools to overcome those difficulties. A SIT program usually involves a combination of different elements, such as sitting on a bouncy ball, being brushed or rubbed with various instruments, wearing a weighted vest, riding a scooter board, being squeezed between exercise pads, and other similar activities.

Is this set of activities helpful in the therapy of disabled children? Unfortunately, we do not have convincing evidence to offer a positive answer to this question. On the contrary, there is extensive proof that SIT could be harmful. Katarzyna, mother of a 12-year-old boy from Warsaw with cerebral palsy named Jas, has had very bad experiences with this kind of therapy: "Rocking my son in a hammock in the SI therapy center wound up with him being rushed to an orthopedist," she recalls. "Nobody bothered to check in advance that Jas has very weak bones."[470] The therapist who applied this treatment probably did not realize that, as far back as in 2005, Tristam Smith, Daniel Mruzek, and Dennis Mozingo from the University of Rochester Medical Center had published an article examining the history of SIT and its uses and misuses; they conclude that most studies indicate that SIT is ineffective, and that its theoretical underpinnings and assessment practices are unproven.[471]

The most exhaustive research review on the effectiveness of SIT was performed by a large group of scientists from the USA, Europe, and New Zealand, in the context of ASD. They analyzed twenty-five studies involving the use of SIT. These are their conclusions:

> Overall, 3 of the reviewed studies suggested that SIT was effective, 8 studies found mixed results, and 14 studies reported no benefits related to SIT. Many of the reviewed studies, including the 3 studies reporting positive results, had serious methodological flaws. Therefore, the current evidence-base does not support the use of SIT in the education and treatment of children with autism spectrum disorders (ASD). Practitioners and agencies serving children with ASD that endeavor, or are mandated, to use research-based, or scientifically-based, interventions should not use SIT outside of carefully controlled research.[472]

And again, if you are inclined to sacrifice your kid's childhood for questionable experiments, I cannot stop you. But you should be aware that they are only experiments, quite often carried out by careless experimenters to boot.

When analyzing such methods as the ones described above, we enter into a strange, paramedical world in which successive weeds of pseudoscience crop up and require thorough exploration and description. The majority of them are assessed negatively by Research Autism. Apart from typically medical procedures, this organization also places daily life therapy and discrete trial training on the list of therapies it feels are unproven or in respect of which the evidence is inconclusive. Holding therapy and social groups have been grouped as "lacking evidence confirming the effectiveness of the method."

Compared to the hydra of pseudoscience, the Lernaean Hydra appears as a zoological curiosity at best worthy of a place in the circus. The hydra of pseudoscience has far more heads than its Lernaen counterpart, and is far more lively since even when you fail to cut off its heads, others continue to grow. A portion of them are threatening for people, just as the venom of the Lernaean Hydra is. The mere work of describing them amounts to an almost Sisyphean task for one individual. Whither the vitality of pseudoscience? I will say more about this in the next chapter.

---

[446] Perseus digital library, "The Lernean Hydra," http://www.perseus.tufts.edu/Herakles/hydra.html
[447] http://www.vojta.com/en/organisation/worldwide
[448] http://www.vojta.co.uk/
[449] http://nydnrehab.com/treatment-methods/dynamic-neuromuscular-stabilization/dns-frequently-asked-questions/

[450] http://forum.gazeta.pl/forum/w,20939,106828906,107403328,Re_szarlatani_lecza_dzieci_.html?v=2
[451] In my earlier book written with Maciej Zatonski *Psychology Gone Wrong* we wrote about the case of a "little Albert." I mean such procedure as described there.
[452] A. Ludewig, and C. Mähler, "Krankengymnastische Frühbehandlung nach Vojta oder nach Bobath: Wie wird die Mutter-Kind-Beziehung beeinflusst?" *Praxis Der Kinderpsychologie Und Kinderpsychiatrie 48*, (1999): 326-339.
[453] http://www.vojta.com/en/the-vojta-principle/vojta-therapy#affix2
[454] K. Stockert, "Akupunktur und Vojta-Therapie bei der infantilen Zerebralparese-ein Vergleich der Wirkungsweisen," *Wiener Medizinische Wochenschrift 148*, (1998): 434-438.
[455] W. Dutka, „Zasady Metody Vojty," Polskie Towarzystwo Terapeutów Metody Vojty, http://vojta.com.pl/index.php?option=com_content&task=view&id=15&Itemid=2
[456] Ibid.
[457] http://nydnrehab.com/treatment-methods/dynamic-neuromuscular-stabilization/dns-frequently-asked-questions/
http://jpgosteopath.com/
http://www.craigliebenson.com/dynamic-neuromuscular-stabilization/
[458] For more, see Witkowski & Zatonski, *Psychology Gone Wrong*.
[459] http://www.researchautism.net/autism_treatments_therapies_intervention.ikml?ra=16&infolevel=4&info=adverseeffects
[460] American Academy of Pediatrics, Committee on Children With Disabilities. "Auditory Integration Training and Facilitated Communication for Autism," *Pediatrics 102*, (1998): 431-433.
[461] M. P. Mostert, "Facilitated Communication Since 1995: A Review of Published Studies," *Journal of Autism and Developmental Disorders, 31* (2001): 287-313.
[462] D. Danielewicz, and E. Pisula, „Kontrowersje Wokół Ułatwionej Komunikacji," in *Wybrane Formy Terapii i Rehabilitacji Osób z Autyzmem*, ed. D. Danielewicz, and E. Pisula, (Kraków: Oficyna Wydawnicza Implus, 2005), 207-224.
[463] O'Donohue and Ferguson, "Evidence-Based Practice."
[464] A. Hall, "'I Screamed, but There was Nothing to Hear': Man Trapped in 23-year 'Coma' Reveals Horror of Being Unable to Tell Doctors he was Conscious," *Mail Online*, (November 23, 2009): http://www.dailymail.co.uk/news/article-1230092/Rom-Houben-Patient-trapped-23-year-coma-conscious-along.html
[465] B. Radford, "Miracle Coma Patient's Story Told via Facilitated Communication," *Skeptical Inquirer 34*, (2010): 9-10; M. Boundary, "Fabricating Communication. The Case of the Belgian Coma Patient," *Skeptical Inquirer 34*, (2010): 12-15.
[466] D. Biklen, *Communication Unbound: How Facilitated Communication is Challenging Traditional Views of Autism and Ability/Disability* (New York: Teachers College Press, 1993).
[467] The discussion took place during the conference: *Pogranicza nauki. Protonauka, paranauka, pseudonauka*, Lublin, Poland November 15-16, 2007.
[468] Y. Sinha, N. Silove, A. Hayen, and K. Williams, "Auditory Integration Training and Other Sound Therapies for Autism Spectrum Disorders (ASD)," *Cochrane Database Systematic Reviews 7*, (2011).

[469] S. Novella, "Auditory Integration Training," *Science Based Medicine*, (June 26, 2013): https://www.sciencebasedmedicine.org/auditory-integration-training/#more-27232

[470] A. Szulc, „Chory Biznes na Chorych Dzieciach," *Przekrój*, (February 16, 2010).

[471] T. Smith, D. W. Mruzek, and D. Mozingo, "Sensory Integration Therapy," in *Controversial Therapies for Developmental Disabilities: Fad, Fashion and Science in Professional Practice*, ed. J. W. Jackobson, R. M. Foxx, and J. A. Mulick, (Mahwah, NJ: Lawrence Erlbaum, 2005), 331-350.

[472] R. Lang, M. O'Reilly, O. Healy, M. Rispoli, H. Lydon, W. Streusand, T. Davis, S. Kang, J. Sigafoos, G. Lancioni, R. Didden, and S. Giesbers, "Sensory Integration Therapy for Autism Spectrum Disorders: A Systematic Review," *Research in Autism Spectrum Disorders 6*, (2012).

Chapter 16: Why is This All so Common?

Parents of children with intellectual disabilities are very tired people. While a quite diverse group, at a certain moment their behavior becomes strikingly uniform. They tell their stories, and when they begin to reach the heart of the matter, they take on a remarkable similarity. It is immaterial whether the speaker is a factory worker, an artist, or a scientist, female or male, rich or poor. None of them are able to comfortably sit upright in their chair. They start to sink down, their heads become too heavy for their necks to remain straight. They are deprived of the energy that lets most of us laugh, shout, or express our opinion forcefully – their voice gradually gets quieter and quieter, as if someone is closing off their air supply. In this awful torpor that emanates from their entire being, there remains something that seems to maintain their flickering fire. This thing is a focus on their own problem. Being so weakened, they become deaf and blind to the majority of things going on around them, which can give the impression of an excessive egoism.

In this state they become particularly susceptible to the strong persuasive messages that are generated by pseudoscience. Planted in such fertile ground, the seeds of pseudoscience grow into thickets of weeds throughout the land of child therapy and rehabilitation, displaying strong resistance to criticism. This is the result of many factors, the majority of which are identical to those that support the massive market in useless therapies for adults. We have discussed them in detail in my previous book, *Psychology Gone Wrong*. To avoid repeating myself, I would only like to briefly recall them here. However, I will go into more detail concerning the particular factors unique to child therapy and rehabilitation.

Writing about factors that serve to maintain the market in worthless therapies, I listed those which I gave the label of internal factors. These were:
- therapists' reluctance for research into the process of therapy and its effectiveness;
- evasion in specifying therapy's objectives;
- proponents of various therapies taking advantage of the placebo effect, time variable, regression to the mean, and similar natural factors as arguments in favor of therapies' effectiveness;
- considering therapy as an art;
- lack of studies with high external validity;
- common acceptance of subjective criteria for evaluation of therapeutic outcomes;
- therapists' perception of the therapeutic process as an opportunity for satisfying their twisted or morbid needs;
- lack of a clear distinction between giving patients pleasure and satisfaction on the one hand, and relieving their pain on the other.

Alongside these factors I also distinguished a range of external factors which lend themselves to keeping ineffective therapies on the market. They were:
- common acceptance of subjective criteria when evaluating therapeutic outcomes;
- a myth, maintained by many therapists, that psychotherapy treats causes and not symptoms;
- the need and the search for an authority figure;
- patients' reluctance to accept responsibility for their own life;
- craving for miracles;
- laziness and ignorance;
- the market conditions;
- psychotherapeutic taboo;
- alleged secrecy of psychotherapy;
- cognitive dissonance;
- fashion.

In writing about the blossoming of psychobusiness, I also analyzed the following elements:
- scientific staffage;
- legal possibilities – absence of sufficient regulations governing therapists in many countries;
- insufficient scientific criticism and reactions by the scholarly community to the expansion of psychobusiness.

The majority of the aforementioned factors are also significant in child therapies, and I refer interested readers to chapter 15 in *Psychology Gone Wrong*. However, the range of mechanisms supporting the emergence and development of pseudosciences in the areas of child psychotherapy and rehabilitation is unique in that we may speak of three sides in the relationship – the therapist, the child, and the parents (caretakers). For obvious reasons, the child rarely participates in the process of commencing and ceasing therapy as well as in assessment of its results, as the child is the primary object of therapeutic and rehabilitative actions. Decisions and assessments remain almost exclusively the domain of the parents and caretakers, and the child has little to say (although some attempt to negate this obvious fact, invoking freedom of contract as justification for the functioning of psychobusiness!). As Wiesław Łukaszewski writes, "Psychobusiness is based on freedom of contract. Those offering psychological assistance (understood as broadly as possible) are on one side, whilst on the other are those who need such help (in the form of therapy, trainings, marketing assistance, etc.)."[473]

Upon learning the diagnosis, in the course of seeking and selecting an appropriate therapy, while carrying out the therapy, in moments of doubt, and in the event of doubts and further searching, there exists a range of factors unique to that situation, while psychological mechanisms with the capacity to significantly disrupt rational functioning come into play. They are deserving of careful study, and we should highlight those of particular significance at individual stages.

Diagnosis

The psychological situation of parents whose child is experiencing a psychological illness or mental handicap is unimaginably difficult. Probably only death or a terminal illness could prove more painful. The emotions that accompany them from the moment the diagnosis is heard are unfamiliar to the majority of us, and even if we make an attempt at describing them, without actually experiencing them we are unable to come sufficiently close to explaining what it is such people actually feel. I have spent many an hour conversing with the parents of children with disabilities and with therapists; no less time have I spent reading stories on the Internet of caretakers who have gone through such tough trials. Thus I am aware that everything I may write about their situation will at best be an approximation of reality, which the majority of them will deem insufficient, and some of them may even take issue with it. However, in desiring to show this world to readers, I am forced to take the risk of some simplifications.

The first reaction of parents in response to a diagnosis is generally an emotional one; indeed, it would be hard for them to behave otherwise. This may come as a feeling of defiance and anger towards the individual issuing the diagnosis: "It can't be true!" As parents, particularly those of children with autism, say, the phase of rejection in the case of a diagnosis issued based solely on analysis of behavioral factors can last for weeks, months, or even years. This is fraught with danger, as the earlier work with the child is begun, the greater the chances are for positive effects.

Even when the parents finally come to terms with the diagnosis, their emotions fail to die down, but rather feelings of self-pity appear: "Why did this have to happen to us?", alongside a feeling of guilt: "We brought this child into the world and we are responsible, so this must be our fault," up to a feeling of determination: "we will do everything it takes to help this child." This mix of the feelings of *guilt*, *failure*, and *determination* is particularly beneficial for pseudotherapy, which will shortly seep into still more episodes of human misfortune. It is sufficient to simply pour a little gas on the fire, skillfully dosing out promises. To create a mechanism that will serve as the fuel of the business for years to come. "'When a very ill child is born, the parents remain moored to life through hope,' explains a mother in *Przekrój*, 'they do not think rationally, and they are in a state of despair, uncertain as to their future and generally left to their own devices. At this moment they are confronted with people bringing ready-made solutions. These people say, "In two or three years at the latest you will have a healthy child." For a fee, naturally.'"[474]

Parents are frequently dismissive of the diagnosis, even though the situation they find themselves in is solely the result of an accident, or the cause of their child's illness is not well known. This is what Jenny McCarthy, the "Playboy" model, was faced with. Upon learning of her son's autism, she was unable to accept the diagnosis and decided to take matters into her own hands. However, because she had, in her own words, "only graduated from the University of Google," she sought answers to her troubles there. In this global trash bin she quickly came across information disseminated by anti-vaccine activists, claiming that autism results from vaccinations: more precisely, by mercury, an element in thimerosal, which itself is supposedly an ingredient in vaccines. In terms of its effects, this is one of the most tragic pseudoscientific statements of our times, and it found an eager apostle in the person of Ms. McCarthy. The conspiracy theory according to which a group of pharmaceutical companies and governments in the majority of countries require vaccinations, and thus create tragedies for thousands of children, was much more to her liking than scientific explanations, or rather the dearth of them. In a world where celebrities (especially those whom nature has given the gift of an alluring physique) enjoy greater authority than professors of leading universities, Jenny McCarthy transformed into an oracle for thousands of parents; particularly after publishing several books, some of which went on to become best-sellers.

I have invoked her story, fascinating from the perspective of the processes involved in media communications while terrifying in other aspects, in order to demonstrate the extent to which the situation of receiving a diagnosis provokes us to seek cause-and-effect explanations. Presently we do not know the exact causes of autism. Most likely, the primary role is played by genetic factors; not ones as obvious as in the case of Down syndrome, but rather distributed throughout the entire genome. Nevertheless, the human mind has evolved in a manner that leads it to seek patterns and cause-and-effect even in places where none exist. We perceive a human face in a random pattern of wood grain, we identify the causes of unfortunate circumstance in the crossing of our path by a cat of a certain color, and the secret of our success in the fact that we came across a coin lying on the pavement. This constant search for patterns and associations was very helpful for our ancestors in the course of evolution. It allowed us to achieve an increasingly quicker understanding of events taking place in the world around us, and in turn to predict them. And yet, in attempting to understand phenomena which we do not understand sufficiently, this property of our mind leads us down blind alleys of simplified, superficial explanations. When we are attempting to understand phenomena that lead to strong emotional engagement – which perfectly describes a situation in which parents are confronted with a diagnosis of their child's illness that upends their lives – the human mind cannot tolerate a vacuum. It is prepared to accept bogus explanations that give the illusion of understanding the situation.

For an average person who is not lost in total ignorance, pseudoscientific explanations do, however, require additional justification. Why are they met with disbelief on the part of thousands of scientists who have frequently devoted their entire life's work to seeking answers to the questions that pseudoscientists are providing them for? Here we are reaching an important point in the process of shaping our understanding of reality. This is the last moment at which we can still step back from the precipice. The fact that thousands of scientific institutions and universities are continually seeking solutions for the treatment of cancer, AIDS, autism, and the other problems that plague mankind, while in the meantime the remedy is identified by someone who has zero authority in those circles can – and at least should – be an argument allowing us to reject the pseudoscientific vision of the world. Yet parents confronted with their child's diagnosis are under the influence of such strong and lasting emotions that they may experience a serious disruption in the process of rational thought. And many of them take the next step, accepting a *conspiracy-based vision of the world*. There is nothing easier than declaring that the majority of scientists involved in research on human health are on the payrolls of Big Pharma. Here is one example of how a father with a mentally impaired child commented on an article presenting therapies for children in a critical light: "And I truly doubt that those methods will ever be considered proper and effective. I doubt that treatment for autism in this annoying country will ever be free and readily available, never mind of respectable quality. And vaccinations? It's all about big money, vaccinations will stay with us forever, everywhere. So I, the maniac, will invest a few thousand more in some worthless method, some awesome experiment that will, like usual, generate positive effects."[475]

Parents can't be entirely blamed for thinking this way. The processes being described are in large measure supported by *the manner in which the health service functions*. Parents encounter tremendous difficulties in obtaining reimbursement for the costs of therapy and in getting access to it. Overworked doctors with contractual limits on patient numbers do not devote much attention to them, frequently treat them brusquely, and mistaken diagnoses are not uncommon. Doctors quite frequently do not have enough knowledge to make the proper diagnosis. When dealing with autism or other difficult cases, a common situation is the one described by Rafał Motriuk, father of a child with autism and also a science journalist employed by Polish Radio and the BBC, who uses his blog to describe the experiences he and other parents go through. Here are two descriptions of visits by mothers to a pediatrician:

> The whole time I was being reassured that they're boys, twins, and premature to boot, that they would catch up to their peers. When they were 3-and-a-half years old I told the pediatrician that Andrzej didn't speak. He shrugged his shoulders and remarked sarcastically "Doesn't speak?", then ended the visit.
> However, when it came time to vaccinate them as 5-year-olds and I again mentioned that Albert wasn't speaking, he blew up and started a fight along the lines of "WHAT KIND OF MOTHER ARE YOU – HOW WILL HE START PRE-SCHOOL NEXT YEAR?", said that I wasn't taking proper care of the

kids, etc. He gave me a referral to a psychologist and said my child had ADHD.
...
"What are you on about? What do they teach you people studying pedagogy... they fill your heads with nonsense about autism, and it's such a rare thing ... Besides, he wrings his hands because he's happy, he doesn't speak and doesn't point because he's distracted, and he puts things in his eye because he's curious about the world" – said the pediatrician.[476]

The duration and quality of contact with the doctor or therapist are key factors in later assessments of the subjective effectiveness of therapy. Studies analyzing such aspects as patient satisfaction from contact with doctors and therapists have shown that in spite of differences in effects, patients participating in fifty-minute short-term therapy sessions were far more satisfied than patients whose meetings with general practitioners lasted not longer than ten minutes.[477] Meanwhile, practitioners of alternative devote great care in listening to their patients, and they spend time inquiring in detail about their patients' medical history. A doctor who is positioned to offer the patient modern medicine's best answers only spends a few minutes in direct contact with them, during which it is not even possible to conduct a full interview.

Because one of the more dominant feelings accompanying parents receiving their child's diagnosis is that of alienation and lack of support, it should be no surprise that they become so attached to sympathetic homeopaths, kind quacks, and other such individuals, without bothering to notice that the manner in which the health service operates and the state of medical knowledge are two entirely different realities which only merge into one whole in their minds. There is no justification in equating the health service with the state of medicine, but on the other hand, conspiracy theories are indeed no great stretch for someone who has gone through bad experiences with the health service (who, after all, has not?) and who has heard of pharmaceutical corporations shelling out their own money inviting doctors to conferences in exclusive hotels, financing the purchase of equipment for their offices, and generally working to corrupt them in other ways. This eminently understandable way of understanding the world, which develops in a most natural manner, serves not only to deliver people into the embrace of pseudoscience. Later on, when doubts arise as to the competence of the first therapist in whom they place their trust, they simply move on to the next, more convincing one remaining outside "the conspiracy."

The only piece of advice which can be given to parents in such a situation is one formulated by Rafał Motriuk in concert with other parents of children with disabilities. It is primarily addressed to parents of children with autism, but it can equally apply to those whose children have been afflicted with another misfortune:

> Yes, your child has autism. It's official and certain. Pretending that it is not so is a natural reaction, something that we all go through, so don't trouble yourself

that you did this or may still be in the process of doing so. Give yourself time to come to terms with the situation. But the sooner you give in to reality, the better for you and for your child. The earlier you accept the diagnosis of autism, the quicker you begin to take action, thereby increasing the chances that your child will achieve independent functioning. Don't be ashamed: at present, 1 in 150 children is born with this disorder. And don't feel guilty, your child's condition is not your fault.[478]

Seeking and selecting a therapy

The manner in which a child's parents or caretakers deal with the news of an autism diagnosis largely determines the selection of therapy. If they are unable to come to terms with a medically justified diagnosis, and have also taken the perspective presented by one of the many conspiracy theories floating around, they have essentially thrown the doors wide open for quacks and cargo cult priests. At the same time, they have also determined the manner in which they will search for a therapist. It will most assuredly rest on *recommendations*, and not those of authorities in a given subject, but rather of parents who have experienced a similar tragedy. They – at least some of them – steered in turn by cognitive dissonance and other mechanisms allowing them to overestimate the validity of their own decisions and behavior, will assume the mantle of super-experts. Here are some examples of how these uncrowned monarchs of misfortune and Internet forums make their case:

> As far as I'm concerned, conventional medicine can label water with sugar a kind of homeopathy. It's helped us quite a bit. This is why I couldn't care less what someone writes and things about all those methods. The most important thing is that they help our children. And the money? We'll repair the holes in our family budget later on.[479]
> Autism is nothing more than possession by evil spirits. Read something by Wanda Prądnicka, who has treated hundreds of such cases. Don't believe if you don't want to, but a little reading won't hurt.[480]
> I know a homeopathic specialist who has cured two children with autism in Słupsk. One girl basically came out of a deep sleep, while in the case of a boy she made contact where nobody else could before. My brother has a stuttering problem, but after two weeks of therapy and medicine his speech has improved substantially.[481]

Some of the interpretations of children's disabilities collected by Rafał Motriuk are even more shocking. They are cited below with comments in the margins:

> "God has a plan, a reason for bringing such children into the world."
> (This is very comforting, especially for atheists and agnostics. Believers, however, begin right away shouting "Eureka!" and singing "Hallelujah").
> "He chose that family because he knew he would thrive there."

> (... it's an established fact that unborn children are wonderful at selecting families, and autistic fetuses are exceptionally skilled in this respect.)
> "He'll grow out of it."
> "It's just an American fad, that autism."
> "The Lord chose you to have him because you're a teacher."
> (Thanks for the information, if I had known earlier I would have been a secretary or a cleaner.)[482]

The technological possibilities for disseminating such opinions have in recent times grown exponentially. Fora and social media portals on the Internet are home to an ever-growing number of discussion groups, and parents of children afflicted with various diseases and syndromes maintain countless blogs. Everything that one can find on the market is given a seal of approval. Indeed, it is quite possible that "parents" praising the merits of particular therapies are in fact the therapists themselves working undercover to promote their products.

In reality, such recommendations should be formulated as such:

> Bartek is on a gluten-free diet. Since he began following this diet, his condition has improved. What can be concluded from this?
> If you conclude that "a gluten-free diet helps treat autism," this would be going too far. Bartek's condition could have improved regardless of his diet. It could have remained unchanged. It could have improved in part because of the diet, and in part of its own volition. There's simply no way for us to be sure. The matter has been under study for over 40 years and nobody has yet come to a definitive conclusion. There are some autism sufferers who choose to avoid gluten because they say it helps them. There are also proper, randomized, double-blind clinical trials with control groups conducted in strict conformance with scientific methods that demonstrate no such link.[483]

It is quite rare to come across such a skeptical interpretation of the effects generated by various treatments; in fact, this kind of statement constitutes an unachieved ideal. There is no counterweight for the mass of unthinking recommendations in the form of reliable sources of information. Despite the efforts of some organizations to educate potential users of their portals on the real causes of diseases and acceptable methods of treatment, finding information based on properly-conducted studies demonstrating the value of particular therapies borders on the impossible.

Statements by parents of children given an incorrect or overly pessimistic diagnosis can be of particularly great weight. An incorrect diagnosis is an event that unfortunately must occur assuming a very large number of cases. In some situations, when the diagnosis is formed on the basis of behavioral indicators in the first years of life, when developmental processes are at their most intense and the speed at which they occur differs among children (a common situation in diagnosing autism), it is a virtual certainty that mistakes will be made. Parents of children issued with an overly pessimistic diagnosis almost invariably turn into the apostles of quacks to whom fate has led them. This comes as no great surprise. When conventional medicine left no room for doubt, the alternative therapist has performed a "miracle"! The mechanism involved in the search for a causal nexus kicks in every time, particularly in these situations; the parents of the misdiagnosed child believe that the pseudotherapy has made a real difference. In this manner, coincidence lends a hand as only by chance could pseudoscience seem effective.

Even in the case of parents who are unapologetic skeptics, pseudoscientific marketing will help them in making their decision. This business is excellent at evoking a *feeling of guilt* arising out of failure to take advantage of missed chances. The conviction that a child's caretakers are under a duty to try all available therapeutic offerings is so deeply rooted that the majority of parents bombarded with recommendations from other caretakers eventually give in.

Another factor contributing to the selection of pseudoscientific therapies is the false belief that *our mind is a blank slate*. We have written of the myriad consequences of this myth in chapter 9 of *Psychology Gone Wrong*. I would like to add here that the dogma of *tabula rasa* also allows us to believe that if something has been written wrongly on the slate, with enough effort it can be erased and replaced with the proper content. Unfortunately, many illnesses involve a genetic component. There has yet to appear the magician who could extract the extraneous 21st chromosome from all the cells of an individual affected by Down syndrome, for this genetic flaw – the presence of an additional chromosome, known as the trisomy – is responsible in around 95% of cases for the emergence of Down syndrome. If the hypotheses are correct that autism is at least partially caused by genetic factors, then they too are irreversible. As I have already stated, it is not possible to erase the contents of the slate and replace them with the right ones, sadly...

In the course of therapy

Writing about psychotherapy in *Psychology Gone Wrong*, we devoted a significant chunk to analysis of how supporters of therapies employ the placebo effect, the variable of time, and the natural tendency of the human body to achieve stasis as explanations for the effectiveness of therapies. In the case of child therapies a similar role is played by naturally occurring *developmental processes*. Regardless of the diagnosis issued to a young patient, whether the child has Down syndrome, autism, a mental disability, or other similar problem, unless it is the victim of a terrible accident or some chemical poising inducing a coma, it will always develop. This is an immanent characteristic of the bodies of young *Homo sapiens sapiens*. Every child will make progress, and our brains will interpret such progress in a manner befitting the situation. If natural biological development occurs during a period when we are sacrificing numerous hours a day on murderously difficult therapy with the child because we have been convinced to do so by some deranged charlatan, we will attribute the results to ourselves, the quack, and the method prescribed by that quack. Here again we see how our autopilot kicks in, seeking cause-and-effect relationships.

If I were sufficiently cynical to conduct such an experiment, I would dream up and begin to employ any ridiculous child therapy, and I am convinced that I would record positive results among my loyal followers. However, I am aware that it would be the result of an extraordinary and powerful force encoded in our genes, the force of development, which marches on regardless of the barriers thrown up along its path. This is the main reason for which children who have been abused with the exercises used in the Doman-Delacato system, caged up in accordance with the recommendations of bonding therapy supporters, or who have spent hours pointlessly drawing figure eights in the air as proscribed by the masters of Ayurveda, succeed in developing. But this development is *in spite of*, and not *due to*, the treatment. If these children had spent time with their loving grandparents, gone fishing with a patient father, accompanied a demanding mother in her household chores, or participated in play with an understanding sibling, the effect could have been far more gripping. However, it could not be called a therapy, and there would be no way to charge a fee; the chances for grants and medals would also be reduced; after all, it is so nice to proudly puff out one's chest to receive an order, never mind that it would best be given to nature…

The strength of nature, which marches on to develop in spite of everything, is the great ally of pseudotherapy. It is enough to place the right glean on interpretations of its effects for parents to remain true to their initial decision. However, it can be the case that in spite of everything the effects are miniscule, and the expectations of parents go unfulfilled. Do they quit the therapy which has failed to justify the hope placed in it? Not a chance! For quacks are prepared to deal with these kinds of situations as well. One excellent illustration is that of Glenn Doman and his therapists. Invoking examples of overly pessimistic or simply incorrect diagnoses, they show what objective can be reached, reinforcing parents' determination. Next, he offers them a program which demands almost superhuman effort. Sooner or later, the execution of such a program must be found wanting, some of the unrealistic requirements must go unmet. This gives the therapist – and the parents! – a perfect answer to the question of why there is no progress. The parents have simply not put in enough effort! A feeling of guilt kicks in, and determination grows. Hitched to this wagon, the parents are incapable of discerning that they are chasing a goal which slips further and further away the more they redouble their efforts. It is like holding a carrot on a stick before the eyes of an ox. No matter how much work the beast does in attempting to reach the carrot, no matter how quickly it runs, the carrot will forever remain the same distance from its nose.

However, during the course of these murderous struggles, something else may emerge, something that most certainly constitutes an effect of the aforementioned child therapies. Setting out down the path, a shared feeling of guilt and a shared objective will unite the parents, cementing a bond between them. As the spiral of guilt begins to take its toll, as greater efforts are made and effects remain imperceptible, the parents begin to view each other with suspicion: "I'm doing everything I can, but what about him? He's stronger, and he could put more effort into it!"; "After all, she's got more free time but spends the same amount of time on rehabilitation for the child as I do!" Similar suspicions, inevitable after years of fruitless labor, lead to the inception of *disintegration of the family*. And if the parents have other, healthy children, it is unusual if those siblings don't begin wondering why the sick one is at the center of their parents' attention? Why can't they live like a normal family rather than everything revolving around their disabled brother or sister? This is all joined by increasing financial demands, as rehabilitation and therapy are not free and someone has to earn money for these things! Money is needed not only to satisfy the day-to-day needs of the family, but also to regularly water the financial tree of the therapy center or the therapist...

"'I ran around from clinic to clinic for years, in search of a ray of hope,' with audible sadness Agata, mother of Kamila, a girl of eight years with autism. 'I left my child in those offices without any oversight. I put masks on my daughter, I stuffed her with junk, because every therapy was supposed to be a means of deliverance. I spent a fortune. I lost two apartments and a husband, who couldn't deal with my stubbornness.'"[484]

This is not an isolated case, something which studies have confirmed:

> Many families affected with autism live in poverty – raising a child with serious developmental disorders is three times as expensive as raising healthy children.
> Families with autism frequently fail to receive necessary support, which leads to enormous stress. Estimates are that even 80 percent of marriages of parents with autistic children end in divorce.
> Only 11 percent of parents/caretakers of autistic children work full-time; 70 percent say that the absence of sufficient assistance/infrastructure makes it impossible for them to remain in work.[485]

Doubts

"I've really been in the dumps recently. After spending a lot of time looking for help, full of hope from reading articles about how autism can be cured, I've run out of strength and I don't know what to do. I just know that I don't know anything."[486]

Someone who has never been in such a situation may certainly pose the question of whether people engaged in such an undertaking will ever come to their senses? Are they incapable of seeing that they're being taken for a ride? The fact of the matter is, they unfortunately cannot. Evolution has equipped us with mechanisms that, in these types of situations, work against us. Alongside the *spiral of the feeling of guilt*, other psychological mechanisms come into play here which reinforce the pointless efforts of parents. In *Psychology Gone Wrong*, in writing about individual therapy we indicated cognitive dissonance as one of the primary mechanisms responsible for cementing a decision once taken. It may also bear a certain responsibility here, if not necessarily the primary one. In the case of child therapy, a far stronger mechanism is the *sunk costs effect*. It arises among stock market investors, but goes far beyond the sphere of economic or professional decision-making, and exerts equal strength in interpersonal relationships. It explains why we are so stubborn (and irrationally so) in our desire to avoid selling increasingly worthless stock or to withdraw from a mistaken investment, why we decide not to change our major when what we are studying fails to interest us, why we remain at a job that gives us no satisfaction, and finally… why we are incapable of ceasing a course of therapy that brings no results. This psychological error results from the widespread psychological trait *aversion to loss*, which, as psychologists' experiments have shown, is even twice as strong as the tendency to seek profit. In order to explain precisely what it consists in, I shall discuss one of many experiments conducted concerning the sunk costs effect.

> For example, Arkes and Blumer (1985, Experiment 2) arranged to have three different types of season tickets sold to persons who approached the Ohio University Theater ticket booth at the beginning of the season. Approximately one third of the patrons purchased season tickets at the full $15 price, one third at $13, and one third at $8. Compared with those who purchased tickets at $15, those who purchased tickets at either of the discounted prices attended fewer plays during the subsequent 6 months. Apparently, those who had "sunk" the most money into the season tickets were most motivated to use the tickets. This is contrary to the maxim that incremental costs and benefits should govern one's decision to attend the plays. Once the tickets had been purchased, all patrons had a license to attend any play. Presumably, the costs and benefits of theater attendance would have been equal for the members of all three groups because participants were assigned randomly to the three price levels. The differential attendance by the discount versus full-price groups was a manifestation of the sunk cost effect: The patrons' sunk cost influenced their attendance decisions.[487]

No great loss if the decisions only affect the frequency of visits to the theater. Michael Shermer illustrates the same effect with far more drastic examples:

> The war in Iraq is now four years old. It has cost more than 3,000 American lives and has run up a tab of $200 million a day, or $73 billion a year, since it began. That's a substantial investment. No wonder most members of Congress from both parties, along with President George W. Bush, believe that we have to "stay the course" and not just "cut and run." As Bush explained in a speech delivered on July 4, 2006, at Fort Bragg, N.C.: "I'm not going to allow the sacrifice

of 2,527 troops who have died in Iraq to be in vain by pulling out before the job is done."

We all make similarly irrational arguments about decisions in our lives: we hang on to losing stocks, unprofitable investments, failing businesses and unsuccessful relationships. If we were rational, we would just compute the odds of succeeding from this point forward and then decide if the investment warrants the potential payoff. But we are not rational--not in love or war or business--and this particular irrationality is what economists call the "sunk-cost fallacy."[488]

Are we therefore still inclined to wonder at parents who plow forward with ineffective therapy, if they have already sunk so much time, energy, and money into it? They are also convinced that just a little more work can bring the desired effect, deceiving themselves in the same manner as the gambler who continues placing bets after wasting a fortune at the casino in the belief that he will recoup his losses.

In addition, the sunk costs effect is reinforced by another mechanism that prevents fishermen from returning to port even when their nets remain empty, that induces warring couples to put off the decision to divorce even though their friends have all written off their marriage, and that allows animals to continue serving us even though we are utterly undeserving of their devotion. This mechanism was discovered by the young, then-future founder of behaviorism, Burrhus F. Skinner. During one particular weekend he realized he had failed to prepare enough food for animals participating in a series of long experiments. As he had no other choice, he decided to reward only every second proper reaction, rather than every proper one as he had done earlier.

He was surprised upon observing that this change did nothing to reduce the speed at which the animals learned. However, the real shock came at the moment when he attempted to extinguish the learned reactions. Extinguishing, which is nothing more than the cessation of rewards, is employed in measuring the strength of the acquired reaction. Researchers count how many times the animal performs a given activity without receiving a reward, until it stops engaging in the previously learned behavior. The more times the reaction appears, the stronger the condition was. In the case of animals "cheated" by Skinner it turned out that their reactions were even more durable than animals rewarded systematically! This was a serious discovery, even though its explanation is quite simple – the individual learns that the reinforcement comes after several reactions without reward. When the extinguishing is commenced, the absence of the reward is nothing unusual, in contrast to an animal systematically rewarded. Only the absence of a reward when it should appear begins to weaken the reaction. In the remaining instances, the absence of reinforcement is entirely normal. This is why Skinner's animals took more time in extinguishing their reactions.

Skinner's discovery initiated an entire series of studies on *irregular reinforcements* and their sequences. As a result, it turned out that the strongest reinforcement of reactions comes from a sequence with diverse time delays. Now you understand, dear reader, why fishermen can sit for hours at a time waiting on a fish to strike at their rod? If the fish swam up to it regularly, let us say, once every hour, then two hours without a hit would mark the end of the session. Meanwhile, fishermen are capable of waiting a very long time for their catch. This is how fish unintentionally teach patience to people.

This is the same mechanism functioning when couples experiencing strife provide each other at various times with positive reinforcement, and then, in the same manner as fishermen, they delude themselves for weeks on end that things really might get better. The situation of a parent careening along the murderous treadmill of an ineffective therapy presents itself identically. From time to time, entirely at random, a chance development, perhaps owed to the course of nature, appears. This is enough for such parents to forge ahead in their delusion, awaiting another such occurrence.

Among the factors shielding us from doubt and hardening our belief in the correctness of our choices, we may also list the *need for exceptionalism*. Are there parents who would not desire for their child to be the beneficiary of some unusually fortunate coincidence? Is there a mother or a father who would not like for their child to become someone extraordinary, exceptional? However, rather than a magically gifted daughter or son, fate has bestowed upon them a child with disabilities, one who requires far greater effort than usual, and that child's successes and achievements will never satisfy the parents' ambition; at least, not during the initial steps down the road of raising a disabled child.

Yet here as well, clever charlatans exploit people's hopes and dreams, selling parents a bill of goods in response to their desires. I again would like to invoke the words of Glenn Doman: "Our individual genetic potential is that of Leonardo, Shakespeare, Mozart, Michelangelo, Edison, and Einstein." "Our individual potential race is not that of our parents or grandparents." "All intelligence is a product of the environment."[489]

This statement refers to everyone, including those with intellectual disabilities, and even those with only one hemisphere of the brain! Does this not fit beautifully in with the expectations of parents? To this we may add tales of the phenomenal talents displayed by idiot-savants, people gifted with exceptional skills in one area (such as music), but otherwise disabled. Many parents of ill children are fascinated by such stories – they imagine that their beloved child with an autism diagnosis can somehow turn out to be a genius. They are undeterred by the brutal statistic that at a maximum only every tenth autistic individual is a savant, and in the case of people with underdeveloped intellectual capacity and/or brain damage, only one in two thousand. As it is, if the problem of autism is a relatively rare one,[490] since one unlucky number has been drawn, why would the next one not turn out to be lucky?

Most frequently, however, the muses of fate do not deign to smile on them, but if they succeed in believing fervently enough in the quacks, they will find their feelings of exceptionality assured. To illustrate how this comes to pass I will again cite the history of Jenny McCarthy, who presently states that she has successfully cured her son of autism with a proper diet and antifungal substances. But this is not the most interesting thing – even more fascinating is that she has become an "Indigo Mom" (from the color of the aura created by the chakra of the third eye), and her son a "Crystal Child." Such children can be identified by the manner in which they begin speaking at a very late stage – due to their strong telepathic capacities, the skill of speech is essentially unnecessary. Isn't this beautiful? Isn't it easier to be an Indigo Mom of a Crystal Child, instead of the mother of an intellectually challenged little boy? The famous words attributed to P.T. Barnum come to mind: "There's a sucker born every minute."

A powerful ally of those quacks working in the area of child therapy, ensuring on the one hand their untouchable status, and on the other hand serving to prevent us from having doubts, is the special sort of *immunity granted to people who help children*. Can we really accuse of cynicism, deception, or ignorance somebody who devotes his entire life to helping children? Who will cast the first stone? It would appear that I have been assigned the thankless task of demasking the true faces of those charlatans, and I seriously doubt many will follow in my footsteps.

Searching further

Let us, however, assume a rational outcome, for in spite of everything such outcomes do occur at times. Parents discouraged by the absence of the expected effects engage in a renewed search, again begin to read through hundreds of conflicting posts on internet forums, and spend long hours browsing through recommendations. Perhaps some of them begin to slowly perceive the delusion perpetuated by the promises on offer, but hope, that irresistible force that nature has gifted us with, does not allow them to rest. There simply must be some way to help our child! We simply cannot accept the heartless scientific perspective that coldly declares there is nothing to be done. There must be a way! It must be found and tried: "I'll say this – if somebody ordered me to stand in a corner facing north-west, look backward through my left arm and spit three times while patting myself on the head, and that this would help my child – I'd do it without a second thought."[491]

Many parents, egged on by pseudotherapists, engage in equally absurd activities. Those who have not yet rejected the conspiracy theories, which is very difficult for someone in their situation to do, return back to the point I have described in this chapter under the title "Seeking and Selecting a Therapy." Many factors contribute to people remaining locked into this cycle. One of them is the activity of media constantly seeking out news and the stupidity of journalists, who can thoughtlessly provide a spark to the flames of hope, with no other purpose than to ensure that their report remains on the front page for a day or two:

"**Autism is curable.**" This is the headline located in the science pages of one of Poland's daily newspapers. My heart leaps for two reasons. The first is obvious. The second?? Well, if something has been accomplished and I am not aware of it, that means I've missed a massive piece of news. But it lasts for only a fraction of a second, because I realize that if a way is discovered "to cure autism," it will not happen overnight. There will be trials with genes, medicine, therapy. Perhaps some genetic flaws can be identified, described, reversed. Or perhaps prevented. Right now, I'm reading the article.

As it turns out, the discussion does not concern the entire autism spectrum, but only Rett syndrome. The symptoms of this disease are similar to those of autism, which is why it is considered part of the spectrum of Autistic Syndrome Disorder, which is also composed of such syndromes as Asperger's.

Aleksandra calls the newspaper and raises hell. After all, they have printed in big bold letters a promise to hundreds of thousands of parents in Poland that their children, previously acknowledged as terminally ill, could now be cured.

The headline on the internet is different from that in the print version: "Autism is treatable." Well, that's nothing new. But the difference between "treating" and "curing" is about the same as the difference between life and death.

Rett syndrome is a serious developmental impairment that only girls suffer from (boys theoretically can suffer from it, but they all die in the womb). It is caused by a flaw in the MECP 2 gene, located on the X chromosome. ...

An article published in *Science* by Prof. Bird of Edinburgh details that he has successfully turned on and off the MECP 2 gene in mice. After turning the gene off, the mice began to display traits associated with Rett syndrome; when the gene was turned on again, these symptoms disappeared and the mice returned to normal health.

I called professor Bird to ask him for some details, and most importantly to discuss the significance of his discovery for other disorders on the autism spectrum. In his opinion, what was important was how previously it was felt that changes at the level of nerve cells were irreversible; now we know that biochemical processes evoked by damage to MECP 2 can be fought against. At least, this is true for mice, and now needs to be tested on people. I ask about the rest of the people on the autism spectrum, as the encyclopedia lists several genes as "leading candidates" for the autism gene, but MECP 2 is not listed among them. In Bird's opinion, damage to MECP 2 is recorded in 1 in 100 autistic cases, but there is no guarantee that only this gene is responsible for the emergence of symptoms, as it is generally considered that if autism is a genetic ailment, the cause is a combination of multiple genetic factors. So that's it for the "curability" of autism. Not in Moscow, but in Leningrad, not Mercedes-Benzes, but bicycles, and they're not giving them away, but stealing them.[492]

The article, however, managed to "exist," draw people's attention; those interested in the subject surely noticed it. Never mind that the hopes of thousands of parents were needlessly raised. The media do not bother to account for this in their cost-benefit analysis. Understanding the real significance of the information also remains the domain of a select few. This is primarily because only a small number of people are curious enough and have the specialist knowledge to understand the real meaning of the false press report. And even if someone jumps through all those hoops, as in this case, the capacity of such an individual to privately reach an audience is incomparably tiny compared to that of a nationwide newspaper.

Remaining trapped in the cursed cycle of seeking new therapies is also heavily influenced by the massive complexity of the therapeutic services market. It allows for a never-ending journey from quack to quack. The number of proposed pseudotherapies is so great that one simply could not find the time to try them all. In the previous section, I referred to a list developed by the specialists at Research Autism which encompasses the one hundred most commonly applied autism therapies. Again – these are only the most frequently used therapies, and only in the case of one illness. If I wanted to name and assess all of those in existence, I would need an army of experts to do so. It is easy to image how much time would be required to try each of them on one's own child, and how costly it would be – for both parent and child.

In addition, clever scammers on the child therapy market sell their services under continually changing names, packaged in ever-evolving jargon. One good illustration of this technique is the manner in which Dennison sells exercises recommended for centuries by practitioners of Ayurveda in a package designed to convince the buyer that it is a method based on the discoveries of modern neurophysiology. This is an old trick.

At times, a chance event can help escape the clutches of pseudoscience, such as in this case:

> I'll just say this – my child was treated with Bobath from birth, when she was 6 we changed to Vojta, and for almost 7 years there was no progress. Rehabilitation didn't just lead to stagnation. My child couldn't stand up on her own, her knees were constantly bent, her feet curled in, and she stood on her toes. The whole time we were advised to avoid surgical intervention, orthopedic assistance and home exercises for general development. At one moment, my daughter dislocated her hip, and it was... the best thing that had ever happened to her. We had no choice, and our daughter went through two operations; one was a severing of the adductor under the knee and the Achilles' heel in both legs, while the other was a "repairing" of the left hip. The rehabilitation that was and continues to be conducted consists in normal exercises for strengthening muscles and improving joint flexibility. There's no name for it, nothing foreign-sounding, no acronyms, but there is tremendous improvement. My daughter is finally walking on her whole feet and can get around on her own. Right now it's just a few steps at a time (not because she can't do more, but she's afraid of falling down), but she's doing fantastic with Nordic walking poles. Because the exercises aren't some secret thing, they can be done every day with my daughter at home, and the therapist even gave me very precise instructions plus a film to watch.[493]

Perhaps the only way of escaping the grip of fruitless searches is the one taken by Agata: "It was just recently, seven years after the birth of her daughter, that Agata stopped visiting the clinics. And she saw Kamila for who she was. She caught a glimpse of an exhausted and terrified child engulfed by terrible, never-ending exertion. And for the first time in so many months, she hugged her. She knew that the loss of her illusions would finally bring her peace, that she could focus at last not on ineffective treatment of her daughter, but simply on her daughter."[494]

---

[473] W. Łukaszewski, "Kto kogo? Komentarz do artykułu Tomasza Witkowskiego i Pawła Fortuny 'O psychobiznesie, tolerancji i odpowiedzialności, czyli strategie czystych uczonych,'" ["By Who to Whom? Commentary on the Article by Tomasz Witkowski and Paweł Fortuna: 'On psycho-business, tolerance and responsibility or strategies employed by pure scientists,'"] *Psychologia Społeczna 4,* (2008): 349–352.

[474] Szulc, „Chory Biznes na Chorych Dzieciach," *Przekrój*, (February 16, 2010).

[475] http://forum.gazeta.pl/forum/w,10034,106667590,106848216,Re_Szarlatani_lecza_dzieci.html?v=2

[476] R. Motriuk, *Autyzm: Dziennik Ojca: Choroba, Terapia, Życie Codzienne,* 62 and 65 Limited printed edition published by the author. An electronic version was deleted from the net on the demand of the child.

[477] K. Friedli, M. King, M. Lloyd, and J. Horder, "Randomised Controlled Assessment of Non-Directive Psychotcherapy Versus General-Practitioner Care," *Lancet 350,* (1997): 1662-1665.

[478] Motriuk, *Autyzm: Dziennik Ojca,* 70-71.

[479] Ibid.

[480] http://f.kafeteria.pl/temat.php?id_p=3994496&start=90

[481] http://f.kafeteria.pl/temat.php?id_p=3994496&start=120

[482] Motriuk, *Autyzm: Dziennik Ojca,* 62.

[483] Ibid, 179.

[484] Szulc, „Chory Biznes."

[485] Motriuk, *Autyzm: Dziennik Ojca,* 101.

[486] http://forum.gazeta.pl/forum/w,10034,106667590,106848216,Re_Szarlatani_lecza_dzieci.html?v=2

[487] H. R. Arkes, and P. Ayton, "The Sunk Cost and Concorde Effects: Are Humans Less Rational Than Lower Animals?" *Psychological Bulletin 125,* (1999): 591-600.

[488] M. Shermer, "Bush's Mistake and Kennedy's Error: Self-Deception Proves Itself to be More Powerful than Deception," *Scientific American,* (14 April, 2007): http://www.scientificamerican.com/article/bushs-mistake-and-kennedys-error/

[489] J. Traub, "Goodbye, dr. Spock: Vignettes from the Brave New World of the Better Baby," *Harpers Magazine,* (March, 1986), 57-64, 58.

[490] In general, the available data is quite inconsistent. Depending on the source, we may cite 1 in 1786 children, or 1 in 111. The relative rarity of the illness may be a consequence of poor diagnostic awareness. On the other hand, alarmingly high results may contain exceptionally pessimistic diagnoses, meaning cases of neurotypical children being diagnosed as autistic.

[491] http://forum.gazeta.pl/forum/w,10034,106667590,106848216,Re_Szarlatani_lecza_dzieci.html?v=2

[492] Motriuk, *Autyzm: Dziennik Ojca,* 34-35

[493] http://forum.gazeta.pl/forum/w,20939,106828906,107403328,Re_szarlatani_lecza_dzieci_.html?v=2
[494] Szulc, „Chory Biznes."

CHAPTER 17: PROTECTING YOURSELF FROM CHARLATANS

That evening I was already weary from a whole day's plugging away at a book. After reading dozens of articles, struggling to keep all of their conclusions straight in my mind, and then attempting to present them to the reader in an accessible manner, I found myself simply unable to muster up a serious mental effort. My thoughts became increasingly shorter and superficial – my brain cried out for rest. I turned the computer off and had a shower. However, to avoid wasting the rest of the day I took *How to Teach Your Baby to Read* to bed with me. Not because I wanted to force myself to engage in a penetrating analysis. I simply wanted to get a feel for the things I was soon going to write about. I began reading. The book was not in the least uninteresting, but just the opposite; I felt as though some substance had been injected into my central nervous system that broke the previous wave of exhaustion. All of this was due to the author of the book. A master storyteller wound a supreme fable, one of a struggle against the impossible. With a narration nothing short of a fairy tale, the potential of overcoming previously unconquerable barriers was so palpable that I simply could not put the book down. The author's unshakeable confidence in his own capabilities in concert with the realism of the description led me directly to the place I was supposed to reach. "Hold on a moment!", I thought, as I began returning to my senses. "Is this really Glenn Doman?" I mean, that surname "Doman." Nevertheless, it's not so unusual as to make confusion impossible. I leapt out of bed, put my robe on, and restarted my computer. I checked the full name from start to finish, finding nobody else, but it is well known that the results of internet search engines are heavily influenced by massive marketing budgets. So I examined the author's photo – without a doubt, the same person. Again I reviewed the articles he had written, checking the dates and locations. It all matched up. Yes, this was the same person – the author of a pseudoscience vision of reality that had seduced so many. It was extremely late when I turned off the computer, and yet for hours I remained unable to fall asleep, thinking about the strength of the message that had so shaken me.

In the process of deepening my explorations of pseudoscientific concepts, I have read many similar books. Because, as a rule, they do not place much great strain on the intellect, I leave them for bedtime reading, and on multiple occasions have found myself swept along by a similar enthusiasm, hope, and certainty of the claims made. I have also frequently been in the position of doubting in my own arguments, at times even checking the next day whether I have not confused last names, whether I was really writing about THAT Simonton or THAT Doman. The persuasive strength of those texts, along with the self-certainty and knowledge of their authors, are so great that my skeptical mind is not infrequently confounded, therefore I check again, and again, and again...

I am in the comfortable situation of not having heard a cancer diagnosis, nor have I been told my child is mentally challenged, nor have I been confronted any of the other misfortunes that serve as fuel for many charlatans. However, when reading some of their works even I could not deny the strength of their message. What, then, must one feel who has experienced such troubles, and who has come across one of these books? Is there any chance to resist their temptation, their magical strength? I doubt it. In the place of such an individual I would probably be unable to myself. This is why the task of answering the question posed in the title of this chapter is such a difficult one. It will be all the harder for someone who is the victim of such a tragedy.

Let us begin by considering the meaning of the word *therapy*. In all of the dictionaries and encyclopedias I have examined, this term is considered an equivalent of *treatment, curing,* or *healing*. For example, *Webster's Encyclopedic Dictionary* defines it as "the treating of the physically or mentally ill by therapeutic means".[495] *Oxford Dictionaries* define it as: "Treatment intended to relieve or heal a disorder."[496] This is nothing strange, as *Online Etymology Dictionary* explains that this word comes to English via the Latin *therapia*, which in turn comes from the Greek θεραπεία, and literally means "curing" or "healing."[497] If we begin to examine the meaning of the word "psychotherapy" we encounter similar definitions, which hold that it is "the treatment of nervous disorders by psychological methods."[498]

So, if therapy is a treatment intended to relieve or heal a disorder, we would expect that it would take a sick person and make him well, take a person with disabilities and restore her full functioning. Thus, I expect therapy to take a child from being autistic to being free of autism, from a diagnosis of Down syndrome to the absence of Down syndrome, from intellectual disability to full intellectual capacity. And if a "therapy" is unable to achieve these aims, perhaps it should not be called a therapy, but rather something else. The question is, what?

Perhaps rehabilitation? Let us again explore the meaning of this word. It is explained as "Restore (someone) to health or normal life by training and therapy after imprisonment, addiction, or illness:"[499] "to bring (someone or something) back to a normal, healthy condition after an illness, injury, drug problem, etc."[500]

In calling the activities that child therapists engage in "rehabilitation," we are certainly closer to what these people in fact do; nevertheless, it remains far from "restore to normal life" or "bringing back to a healthy condition." We would be more inclined to agree with the statement that their activity is intended to restore the maximum possible level of physical and/or mental functioning.

When speaking of children with Down syndrome, autistic children, or those otherwise intellectually impaired, we should not use the term "therapy," nor even "rehabilitation", but rather *education*. In reality, therapists are spending the majority of their time *educating* the children. They teach them to speak, to use objects important in daily life, to function in the world around them, to understand reality, and many other things. This is a special kind of education that requires special skills, and demands far greater effort than what is expended on the education of those who are untouched by such problems; nevertheless, it remains simply education. In fact, if we define what therapists do in this manner, it occurs that those who teach and educate children with disabilities generate the best "therapeutic" effects. But they *do not cure* children. They teach them many useful and important things, they rehabilitate, and they attempt to restore normal functioning.

Goodness knows why people would prefer to be therapists rather than solid physiotherapists or teachers. Perhaps "therapist" is a more prestigious profession? One can search near and far without finding educations of intellectually disabled children, while "therapists" are a dime a dozen.

Several years ago I attended to a scientific conference devoted to therapy of disabled children. With growing disbelief I listened to presentation after presentation which taught me that in my childhood I had been subjected to not less than a few dozen types of therapy! To this day, my parents remain unaware that they had put me through so many wringers. I don't think I'll even bother to try explaining it to them, because if I did, either I or they would end up actually needing therapy.

My home was always graced with the presence of a dog, and my siblings and I played with it. We treated this as a natural part of life. During the conference I found out that I was being subjected to *dog therapy*. Perhaps this is why I can write books, as opposed to some of my classmates who didn't have dogs and are yet to publish anything? This is not all. Imagine that my parents, likely blessed with uncommon therapeutic intuition, allowed me to play in the yard with other children, surrounded by stimuli, and this was their way of carrying out the postulates of *intensive sensory stimulation*. During summer vacations in the countryside I was subjected to *equine-assisted therapy*, and for a short time in my life, until my selfish guinea pig ran away, I was under the influence of *guinea pig therapy*.

I could go on much longer about my adventures with various "therapies." Or, rather, I could simply parrot the statements of their supports reciting the real and the imaginary effects of assorted aspects in the reality that surrounds us on our well-being, health, psychological wellness, physical condition, and intellectual prowess. However, this is not the objective of the present volume; one may visit any number of pedagogical conferences to listen to speakers on the subject. After several such conferences and countless lectures, I am unable to resist the conclusion that the "therapists" have chopped the world surrounding us into pieces, given each of them a proper label, and put on their best "smart faces" so as to traffic in everything that was once the primary domain of children, playing a key role in their development. They haven't even bothered to put in a lot of effort coming up with names. It is sufficient to take the name of some object or an animal from our environment and add the suffix "-therapy." The list that appears below is not the work of my imagination. I simply entered into Google what came into my mind. As it turned out, I was right – essentially everything can be made into a therapy. All of the items on this list are available for purchase at the market of health and well-being:

- color therapy (or chromotherapy),
- flower therapy,
- floral thrapy,
- sound therapy,
- touch healing therapy,
- aquatic therapy,
- stone therapy,
- sand play therapy,
- sand tray therapy,
- nature walk therapy,
- walk and talk therapy,
- laughing therapy.

That is only a taste.

Of course, I agree with the general proposition that all factors which may support development and improve self-esteem are worthy of attention and deserve my recommendation. But must we call them "therapy" without hesitation?

My advice for parents seeking help for their children is thus: if your child is afflicted with a treatable illness (curable in the opinion of modern medicine, not of charlatans), seek a therapist and begin treatment. In respect of psychological disorders, I would refer you to Chapter 10 of this book. The procedure described there for reviewing a therapist and a therapy are equally applicable in respect of children.

If your child has been diagnosed with autism, Down syndrome, or any other disorder that modern medicine and psychology regards as incurable, get another opinion to confirm this diagnosis and seek an educator who will aid the child in achieving the maximum possible level of functioning. Reliable professionals will not promise miracles, and when asked about the nature of the therapy (unfortunately, this is the label many of them apply to what they do) they will readily admit that their efforts generate educational and behavioral effects rather than treatment. Beware of therapists that promise miracles, and do not believe those who recite conspiracy theories coupled with criticism of modern medicine and science as petrified and stuck in the past.

If, in the course of seeking assistance for our child, we come across "therapists" employing methods and exercises that they do not wish to disclose (except for an additional fee), and if everything they do has been given intelligent-sounding names, this is your opportunity to take the right decision and head for the hills as quickly as possible. You can rest assured that you have almost certainly encountered yet another charlatan. Proceed in the same manner with all those attempting to sell you vitamins and supplements from other distant places at massively inflated prices. The majority of these "miraculous" medications can be purchased at your local pharmacy. Yes, that one just down the street...

Sadly, public institutions offering or financing treatment are not always deserving of your trust when it comes to their assessment of a particular therapy's value. As I have already written, some of them support, or even finance, quackery in child therapy, causing a great deal of harm. We have become accustomed to treating the state as a reliable source of knowledge. Unfortunately, this trust in respect of the issue discussed herein is misplaced, and the fact that a state institution promotes a given therapy does not provide us any information as to its effectiveness.

Therapeutic services addressed to children are often associated with the necessity of bearing pain inflicted by a loved one, as well as with intense physical or intellectual effort. Frequently it is suggested to engage in multiple courses of treatment at the same time, following the principle "more therapies means better results." While there is absolutely no evidence to back up this ridiculous statement, it is all too common to encounter children whose daily schedules are full of similar activities. Before we implement such a regime for our child, we must ask ourselves the question of whether we ourselves could handle such a burden. For how long? Would we like to be treated in this way? Would we select such a therapy for ourselves?

It seems that among the advice which can help protect you from quackery, one thing is of the greatest import – the therapy being sought out *is for the child, not its parents*. Sadly, the majority of parents refuse to understand and accept this. For this reason, their heroic efforts amount to nothing more than an attempt at satisfying their own ambitions. They hitch their child to this wagon and are often proud of their "accomplishments," which the child neither understands nor needs. I have seen extraordinary feats performed by children with intellectual disabilities. In these situations I have always posed the question to myself of whether they were the accomplishments of the children, or rather of the parents and therapists? The answer was usually that they were the accomplishments of the parents, and what is worse, created for parents who, lost in their own imaginations, have forgotten that children are not created for them but are rather separate human entities who are often quite different from us. We should help them to the greatest possible extent in their efforts to achieve maximum mental capacity, but we must not apply our imagination as the metric.

I would like to conclude by citing Ewa Pislua's list of the most important criteria in the selection of therapies for children, which serves as a complement to my divagations and should assist in assessing proposed courses of treatment. I am in full agreement with them.

- **Certainty that the method does not cause harm.** Acting to the detriment of the child means not only the application of methods that worsen its health or cause suffering. Treatment associated with unnecessary burdens that does not lead to improvement is also harmful, as it limits the potential for conducting more effective activities.
- **Solid theoretical grounding,** justifying the sense of undertaking certain activities (i.e. providing an answer to the question of why such actions will be helpful). They should account for the sources of problems. Clearly formulated objectives should result from them, which will be pursued on the basis of the proposed therapeutic undertakings.
- **There is the potential to conduct empirical verification of the effectiveness of a given method.** It is impermissible to apply methods which assume that any attempt to examine their effectiveness is doomed to failure, as it would damage the trust that the child must feel towards the therapist.
- **Confirmation in empirical and clinical data of the method's effectiveness.** Verification should be performed in accordance with scientific rigor. The results of such experiments should be published in refereed scientific journals.
- **Development of standards for the application of the method** which show precisely in what manner the method should be applied.
- **Awareness of the method's limitations.** This means that it is possible to clearly state in which circumstances the effectiveness of the method will be poor. There are no therapeutic methods which are appropriate for all problems experienced by children in their development.

It is of particular importance to document the course of the therapy, to perform an objective measurement of the effects, and to examine effectiveness in a manner consistent with the principles of scientific method.[501]

---

[495] *The New Lexicon Webster's Encyclopedic Dictionary of the English Language. Canadian Edition* (New York: Lexicon Publications, 1988).
[496] http://www.oxforddictionaries.com/definition/english/therapy
[497] http://www.etymonline.com/index.php?allowed_in_frame=0&search=therapy&searchmode=none
[498] http://www.collinsdictionary.com/dictionary/english/psychotherapy
[499] http://www.oxforddictionaries.com/definition/english/rehabilitate?q=rehabilitation#rehabilitate__10
[500] http://www.merriam-webster.com/dictionary/rehabilitate
[501] Pisula, „Kontrowersje Wokół Stosowania."

Dear Richard

I hope you do not mind me addressing you that way. You caused quite a stir with that speech of yours in Caltech in 1974, and I have never brought myself to tell you that. I know your intentions were good, but whether it was due to your nonchalance, or to a lack of preparation, you caused the social sciences to become a synonym for a cargo cult in people's minds. You tried to present good and bad practices, and probably meant to elucidate the differences between them rather than to express loathing towards the social sciences as such. However, most people tend to simplify things and, unfortunately, the label "cargo cult" has stuck to us. I am writing to us because I am a psychologist. To those of us who try to engage conscientiously in what they do – as was the case with Young and the rats, whom you talked about in Caltech – it makes many things difficult. Physicists, chemists, and naturalists in general do not take us seriously. Some of them do not even know that psychology is also a science. Can you imagine how difficult that makes it to carry out interdisciplinary projects? Those from our community who indeed practice cargo cult science stand their ground and fight for recognition of their work. This only makes matters worse.

I have accused you of a lack of preparation. I am prepared to justify myself. On the one hand, my words were the result of what you said at the end of your speech: "May I also give you one last bit of advice: Never say that you'll give a talk unless you know clearly what you're going to talk about and more or less what you're going to say."[502]

I am sure that this was not just for show. I have read your speech a number of times, and although I agree with many of the theses presented, I do feel it was chaotic. Examples of bad practices in science came to your mind far easier than good ones. We, psychologists, call it accessibility of cognitive categories. I also find it much easier to generate such examples, but it's not exactly ... fair. It's not entirely in accordance with what you encouraged the graduates to do. Apart from the examples of studies on Young's rats and experiments regarding parapsychological properties carried out by Rhine, you presented nothing developed by the social sciences that works and is effective. Today I would like to convince you that there are many such things, that the social sciences display immense potential, if only scientists adhere to your recommendations. And there are many who do so. Allow me to tell you about some of them.

You could not have learned of the first one when you were giving your speech, since it began at the same time as you were speaking before the students. It was then that the article presenting the first study results was published. It is all the more interesting considering that the central figure in this story, John Voevodsky, was also working in California at the time. He has and continues to contribute to saving the lives and health of millions of people. What is more, he protects us from unnecessary spending – all due to his thorough research. As early as in the 1960's he measured the reaction times of drivers to changing brake lights in cars, and the possibility of reducing the probability of accidents. His research was carried out, as you described in Caltech, honestly, with all possible options taken into consideration. At the beginning of the 1970's he began to experiment with an additional, third, brake light, which in his experiments was yellow and flashing. In order to verify if he was right that it would decrease the probability of accidents, he equipped 343 San Francisco taxicabs with the third brake light and left 160 taxis with no additional light as a control group. These taxicabs were assigned to drivers at random, regardless of their preferences. Over 10 months the taxicabs from the experimental group covered 12,300,000 miles, while the cars from the control group did 7,200,000 miles. What were the results? Taxis with a third brake light suffered 60.6% fewer rear-end collisions than the control-group taxis. Additionally, drivers of taxis with the third brake light that were struck in the rear by other vehicles were injured 61.1% less often than were drivers of taxis without the light, and repairs to all taxis with the light cost 62.8% less than repairs to taxis without the light.[503] Amazing, is it not?

I know you will ask me to replicate these results, to examine their limitations, etc. This has all, naturally, already been done. The National Highway Traffic Safety Administration (NHTSA) repeated Voevodsky's studies on a much larger scale and reached the same conclusions. As a result, since 1986 passenger cars, and since 1994 also light trucks, are required to be equipped with an additional brake light. In order to check the effectiveness of this light, the NHTSA has analyzed police reports from eight states and found that it reduced the probability of rear-end collision by 4.3%. Although this result is far less striking than those of Voevodsky's experiment, it still proves that since the third brake light has become mandatory, in the USA alone there have been 200,000 fewer collisions, 60,000 fewer injuries, and more than 600,000,000 dollars that would have been previously spent on liquidating the damage caused by accidents were saved. In other words, each dollar spent on the production and installation of a third brake light brought 3.18 dollars of savings.[504]

Do you think that such research and its results can and should be labelled a cargo cult? Probably every car manufactured today is equipped with an additional brake light, and nobody even thinks about who invented it or wonders about its consequences. What is more, Voevodsky's name remains virtually unknown. He is not discussed in psychology textbooks, although they are full of characters I would not hesitate to call priests of the cargo cult.

This is not an isolated case of such magnificent results being generated by psychological studies. Equally interesting are the studies that have led to the majority of today's emergency vehicles (ambulances, fire engines, etc.)

being equipped with visibility vests, and some protective clothing being yellow-green. Why? Psychological studies of human visual and auditory perception show that because the color-transmitting cones in our eyes don't work well in the dark, some colors are easier for us to see at night. We are most sensitive to greenish-yellow colors under dim conditions, making lime shades easiest to see in low lighting.

These regularities were known to two researchers who, coincidentally, were members of a volunteer fire department – Stephen S. Solomon and James G. King. They analyzed data on accidents in Dallas involving fire engines. It so happened that in the 1970's and 80's the Dallas Fire Department had begun changing red fire engines for yellow-green ones, and later bought new red vehicles with white cabins. The researchers analyzed accidents from a four-year period, taking into consideration only those involving two or more vehicles. It was found that the risk of an accident caused by low visibility can be even three times greater for red and red-white vehicles in comparison with yellow-green ones! It was also found that the risk of injury sustained during such accidents was far lower if the accident involved a yellow-green vehicle.[505] This analysis only confirmed earlier observations made by Solomon during analysis of 750,000 Fire Department interventions in nine US cities. Yellow-green vehicles were involved twice less frequently.[506]

In this case as well it is difficult to overestimate the gravity of this simple finding. One can only estimate the number of human lives saved from death or handicap. Is it not a wonderful result? Is it not true that, in comparison to this result, the sometimes lifelong attempts at proving some trace effectiveness of new psychotherapy seem pathetic? You will probably agree that research into preventing and limiting accidents is, compared to studies into methods of coping with trauma (e.g. after an accident) much more sensible? In this case as well, it is immensely unfair to label such work cargo cult.

The priests of the cargo cult are employed in guiding airplanes to the ground, or rather mimicking such actions. You will be surprised to hear that psychologists have helped bring numerous planes safely to ground. At this very moment, thanks to the efforts of psychologists, 999,996 of every million planes taking off in civilian aviation worldwide will land safely, while at the beginning of the 1960s a few dozen of planes per million that took off were involved in accidents.[507]

It was due to such a high accident rate in aviation that the causes of the crashes began to be investigated more thoroughly. Researchers quickly concluded that in over 70% of aviation accidents the cause was human error, rather than equipment malfunction or weather. While examining the effects of human error in aviation, NASA found that the majority of them were caused by problems with leadership, coordination of teamwork, and decision making.[508]

In order to solve these problems, two psychologists, John Lauber and Robert Helmreich, were employed to design a new type of psychological training for crew members. They concentrated mainly on group dynamics, leadership processes, communication, and decision making. This training is presently known as Crew Resource Management (CRM).[509] Additionally, Helmreich and his associates designed a method called Line Operations Safety Audit (LOSA), used to monitor the effectiveness of CRM regarding risk management of human error.[510]

As recommended by the United Nations, CRM was introduced as a mandatory element of air crew training for civilian airplanes in 185 countries. It is difficult to estimate how many human lives and material goods have been saved by this program. Data obtained in LOSA research shows that its positive influence is immense. In 98% of all flights, the crew encounters some kind of danger at least once, if not more frequently. There is an average of four dangerous situations per flight. Errors were observed in 82% of all flights. There was an average of 2.8 errors per flight. Most of them are handled by the crew without any additional consequences, mainly due to the practices entrenched by CRM.

The success of this method encouraged similar practices to be used in perfecting teamwork coordination. At present, studies are being carried out regarding the use of this form of training in the work of medical teams (mainly those working in operating theaters or in rescue missions).[511] CRM is also used in industrial conditions, e.g. on oil rigs,[512] in nuclear power plants, and every other place where it is vital to act flawlessly and quickly in the face of imminent danger.

The author of yet another, strikingly simple, use of findings from behavioral psychology is Bart Weetjens, an industrial engineer who abandoned his profession in order to commit himself to helping solve problems faced by people in Africa. In 2010, while partaking in a TED conference, he described his fascinating work. Using the standards of operant conditioning analyzed by behaviorists, he taught rats to detect landmines. This is a major problem in countries where wars have taken place. Weetjens reports that in the space of just one year approximately 6,000 people step on landmines. Most of them die or are left disabled. Thanks to the implementation of Weetjens' ideas in Mozambique, it was possible to train rats for approximately 1/5 of the cost of training a mine-detecting dog. It was also possible to reduce the cost of demining of one square meter from 2 dollars to 1.18.

However, this is quite a minor thing compared to the other ideas of this rodent enthusiast, as he calls himself. A real problem that a large number of people in developing countries is faced with is tuberculosis. Just in the year preceding Weetjens' lecture, nearly 1,900,000 people died due from this illness as the main infectious agent. Especially in Africa, where tuberculosis often coincides with HIV, it is an immense problem. Early detection of the illness could at least partially help solve this problem. Unfortunately, the standard germ identification procedure – by microscopic sample observation, per WHO recommendations – ensures only 40 to 60% accuracy. In Tanzania, tuberculosis is detected in only 45% of infected patients before they die. This grim picture could change if it were possible to detect it earlier and with more precision. This is where Bart Weetjens and his rats, like Harry Potter equipped with magical powers, step onto the stage filled with human tragedy.

Trained by their master to use the sense of smell, the animals recognize infected samples with an accuracy of 89% and specificity of 86%. What is more, the animal needs approximately two seconds to do this. Therefore, it is possible to analyze in seven minutes the same number of samples for which an analyst would need an entire day of lab work. For example, in five clinics in Dar es Salaam, Tanzania, with a population of 500,000, 15,000 of them came in order to be examined. By simply presenting the samples to the rats and undertaking later analysis of microscopic tests, it was possible to increase the detection rate by over 30%. This is a remarkable result, considering that a patient with tuberculosis going undetected during microscopic analysis can infect up to 15 healthy people per year. Weetjens' rats have saved many human lives.

This contemporary Pied Piper of Hamelin, when talking about the capabilities of his favorite animals, describes their limitless possibilities: "We're talking now explosives, tuberculosis, but can you imagine, you can actually put anything under there. ... Can you imagine the potential offspring applications – environmental detection of pollutants in soils, customs and applications, detection of illicit goods in containers and so on."[513]

At the moment, Bart Weetjens proudly presents his animals with cameras attached to their backs. He teaches them to find victims of earthquakes, construction disasters, and other similar tragedies. Is this not genius at work? And it is only a simple combination of the animals' natural talents and their ability to learn, accurately described by behaviorists.

Other examples of effective applications of psychology are:

- Behavior-Based Safety programs which help companies cut accidents and injuries through systematic observation, analysis and intervention.[514]
- Psychological testing which helped design the omnipresent push-button keypad for the telephone. The design is the best in terms of performance. Again, we should look at this discovery from the point of view of people who are in an emergency when every second counts.[515]
- "Psychological research shows that comprehensive sex education and HIV prevention programs are effective in reducing high-risk sexual behavior in adolescents. Based on over 15 years of research, the evidence shows that behavioral intervention programs that promote appropriate condom use and teach sexual communication skills reduce risky behavior and also delay the onset of sexual intercourse."[516]

I will ask the same question again – don't you think that it is offensive to call the work of all these people cargo cult? I could fill many pages by writing about similar, specific examples of the social sciences at work.

I do not know if you are aware, but when I speak during my lectures about similar achievements I am fascinated by, and whose authors I view as heroes, I am faced with accusations of reducing the meaning of psychology to things as prosaic and common as additional brake lights in a car. Can you imagine that? Engaging thousands of psychologists to hold hands and comfort people who survived accidents, catastrophes, and similar events is something perfectly appropriate, something called the "spiritual" sphere of action, but when someone saves the lives and health of thousands of people, it is considered coarse. When I first heard such accusations I was bitterly surprised. Now I have grown accustomed. In order to complement the picture of psychology as an effective branch of science, I will tell you about a somewhat more complex achievement which, by some people, as my critics, may be viewed as a little more "spiritual." It will still, however, be far from "the spiritually deepest" searches of codependences between id, ego, superego, and other such activities which you have rightfully termed the cargo cult.

Imagine that in recent years a group of persistent researchers has begun to successfully measure and compare what for decades seemed unmeasurable – the effectiveness of psychotherapy. This was possible by mimicking the newest trends in studying the effectiveness of treatment in medicine. This time, however, the mimicking does not stop at the stage which believers of the cargo cult reached. At present, the study of psychotherapy corresponds with the traditions of Evidence Based Medicine (EBM). The pioneer of this approach is considered to be Ignaz Semmelweis, who, at the end of the nineteenth century, while analyzing statistical data, noticed a relation between female mortality in maternity wards and the previous work of doctors in the dissection room. Without knowing the real causes of postnatal infections, he discovered a method of preventing them. Medicine based on facts or evidence, as the phrase is commonly translated, developed in part as a result of the helplessness of doctors searching for the causes of illnesses and methods of treating them, and in part as a result of the availability of continually expanding resources of epidemiological data and results of experimental studies.

In my opinion, some doctors were in fact using EBM much, much earlier. One such precursor of EBM was an English doctor, Edward Jenner, who introduced smallpox inoculation at the end of the eighteenth century. In his research he applied a certain folk wisdom which says that women who milk cows do not get infected with smallpox. These women most often contracted a very benign type of zoonotic disease - cowpox. Therefore, he came to the conclusion that suffering from cowpox inoculates people against more severe types of pox, and proposed purposefully infecting people with this illness in order to protect them from smallpox. Jenner was obviously faced with ridicule from the scientific community, which, without knowing the mechanism of the illness, could not believe such hypotheses. However, the researcher carried out his own "inoculations" in spite of their mockery, and proved that these people were immune to smallpox. The procedure used by Jenner describes, in greatest simplification, the rules of EBM: "Observe what works, compare what works, and then use it. Leave understanding of these processes for later."

In psychotherapy we talk about Evidence Based Psychotherapy (EBP), but the working mechanism was adapted from medical practitioners – we observe what brings results, compare various methods of operation, and then we apply and perfect them. In recent years, several expansive monographs have been published presenting in detail the basics of measuring the effects of psychotherapy and the rules for monitoring influences based on dependable foundations.[517]

You are probably wondering how this could all have been achieved. How come something that used to be called art has become a reproducible and measurable procedure? Mainly due to scientific rigor. Therapy included in EBP research is carried out on the basis of a therapeutical textbook, which provides it with reproducibility. In order for research to be included in EBP comparisons, it must meet certain standards, such as: the necessity to carry out at least two well-planned experiments presenting the differences between groups in which control or comparison groups receive medicine, are subjected to placebo therapy, or other types of therapy. Furthermore, the obtained results should come from at least two independent researchers or research teams.[518] Please note that such standardization of methods facilitates not only the measuring process, but also aids in teaching future therapists. Algorithms of conduct presented in textbooks tell the therapist what to do, and research results suggest the forms of influence to choose.

Thanks to EBP research we have identified leaders among the types of therapy – behavioral, cognitive, and cognitive-behavioral therapy. Their success probably comes from the fact that they were created by scientists focused during therapy implementation on an accurately formulated and measurable target, measurement, and effectiveness. What is more, these are some of a small number of therapies which have detailed textbooks at their disposal, and this means that every person using a given method will take similar actions, in a given order. In most studies and reports, they rank highest.[519] Through research, they can be constantly improved, which in consequence brings even better results.[520]

You will probably ask if this has led to the disappearance of all those strange therapies you encountered and talked about in Caltech. Unfortunately, the answer is no. There are even more of them, and in the preceding sections of this book I have attempted to explain the reasons why, in spite of the good and proven methods at our disposal, we choose solutions resembling a cargo cult more than science. It was important to me to show you that among the practitioners of the social sciences, especially psychologists, there are many who approach their profession with honesty, and do work in respect of which the term cargo cult is highly offensive.

In any way, the *evidence based* approach is presently entering newer and newer regions of social sciences. There is a strong tendency to develop education based on evidence (*evidence based education*), which may solve the teaching problems you discussed in Caltech. *Evidence based management, evidence based training, evidence based leadership*, etc. are discussed more and more often. It seems that a new trend is emerging in the social sciences: to base one's practices on evidence, a trend for what is scientific and empirical. Obviously, as with every trend, it will probably entail numerous simplifications, along with errors in the understanding and utilization of this approach. However, it is a far more advantageous trend compared to one which dictates a retreat from empiricism in the direction of relativism and postmodernism.

Perhaps less spectacular, but no less important, are the achievements of psychology and social sciences in the field of understanding certain phenomena and forming coherent explanatory theories. Their effects are not always precisely measurable, but no less impressive. Daniel Kahneman and Amos Tversky have carried out an extensive amount of research in order to understand decision-making processes in people experiencing a state of uncertainty. As a result, they formed a theory of perspective which is contradictory to the theory of expected utility currently dominant in the mainstream of economic and decision analysis. However, their concept is far more accurate at predicting the behaviors and economic decisions people take. For their use of psychological tools in economic studies, with special consideration of the perspective theory, Kahneman and Vernon L. Smith were awarded the Nobel Memorial Prize in Economic Sciences in 2002 (commonly known as the Noble prize in economics). Due to such researchers as Tversky, Kahneman, Smith, and many others, less known and not honored with similar awards, we are able to better understand human behavior, and those of us who so desire are able to employ this knowledge in solving real social and psychological problems.

Dear Richard, I hope that the information I have provided will lead you to rethink and review your assessment of the social sciences.

Kind regards,
Tomasz Witkowski

Wrocław, 31st January, 2016.

[502] Feynman, "Cargo Cult Science."
[503] J. Voevodsky, "Evaluation of a Deceleration Warning Light for Reducing Rear-end Automobile Collisions," *Journal of Applied Psychology 59, (1974)*: 270-273.
[504] http://www.apa.org/research/action/brake.aspx
[505] S. S. Solomon, and J. G. King, "Influence of Color on Fire Vehicle Accidents," *Journal of Safety Research 26*, (1995): 41-48.
[506] S. S. Solomon, "Lime-Yellow Color as Related to Reduction of Serious Fire Apparatus Accidents: The Case for Visibility in Emergency Vehicle Accident Avoidance," *Journal of the American Optometric Association 61*, (1990): 827-831.
[507] Naturally, these numbers vary depending on the year, on the fact whether we include private planes, or only those with jet, or even propeller propulsion, etc. Data from: http://www.boeing.com/news/techissues/pdf/statsum.pdf
[508] G. E. Cooper, M. D. White, and J. K. Lauber, ed., *Resource Management on the Flightdeck: Proceedings of a NASA/Industry Workshop (NASA CP-2120)*. (Moffett Field: NASA-Ames Research Center, 1980).
[509] R. L. Helmreich, and H. C. Foushee, "Why Crew Resource Management? Empirical and Theoretical Bases of Human Factors Training in Aviation," in *Cockpit Resource Management, ed.* E. Weiner, B. Kanki, and R. Helmreich, (San Diego: Academic Press, 1993), 3-45.
[510] R. L. Helmreich, J. R. Klinect, J. A.Wilhelm, B. Tesmer, D. Gunther, R. Thomas, C. Romeo, R Sumwalt, and D. Maurino, *Line Operations Safety Audit (LOSA)* (Montreal: International Civil Aviation Organization, 2002).
[511] R. L. Helmreich, "Managing Threat and Error to Increase Safety in Medicine," in. *Teaming up. Components of Safety Under High Risk, ed.* R. Dietrich, and K. Jochum, (Aldershot: Ashgate, 2004).
[512] R. H. Flin, "Crew Resource Management for Teams in the Offshore Oil Industry," *Team Performance Management 3*, (1997): 121-129.
[513] http://www.ted.com/talks/lang/en/bart_weetjens_how_i_taught_rats_to_sniff_out_land_mines.html
[514] E. S. Geller, "Behavior-based safety in industry: Realizing the Large-Scale Potential of Psychology to Promote Human Welfare," *Applied & Preventive Psychology 10*, (2001): 87-105.
[515] R. L. Deininger, "Human Factors Engineering Studies on the Design and use of Pushbutton Telephone Sets, *Bell System Technical Journal 39*, (1960): 995-1012.
[516] American Psychological Association, "Risky Business: Curbing Adolescent Sexual Behaviours with Interventions," *Science in Action*, (2006), http://www.apa.org/research/action/risky.aspx
[517] C. D. Goodheart, A. E. Kazdin, and R. J. Sternberg, ed., *Evidence-Based Psychotherapy: Where Practice and Research Meet.* (Washington DC: American Psychological Association, 2006); A. E. Kazdin, Evidence-Based Treatment and Practice. New Opportunities to Bridge Clinical Research and Practice, Enhance the Knowledge Base, and Improve Patient Care," *American Psychologist 63*, (2008): 146-159; L. Luborsky, and E. Luborsky, *Research and Psychotherapy: The Vital Link* (Lanham: Jason Aronson, 2006).
[518] For more, see: D. L. Chambless, M. J. Baker, D. H. Baucom, L. E. Beutler, K. S. Calhoun, A. Daiuto, R. DeRubeis, et al., "Update on Empirically Validated Therapies, II," *Clinical Psychologist 51*, (1998): 3-6.

[519] Task Force on Promotion and Dissemination of Psychological Interventions, Division of Clinical Psychology, American Psychological Association, "Training in and Dissemination of Empirically Validated Psychological Treatments: Report and Recommendations," *The Clinical Psychologist 48*, (1995): 3-23.

[520] Chambless et al., "Update on Empirically"; A. D. Reisner, "The Common Factors, Empirically Validated Treatments, and Recovery Models of Therapeutic Change," *Psychological Record 55*, (2005): 377-399; W. C. Sanderson and S. Woody, *Manuals for Empirically Validated Treatments: A Project of the Task Force on Psychological Interventions* (Oklahoma City: American Psychological Association, Division of Clinical Psychology, 1995).

CPSIA information can be obtained at www.ICGtesting.com
Printed in the USA
BVOW06s1743110916

461809BV00024B/210/P

9 781627 346092